Bart Taylor
'46

GAYLORD

Audubon Perspectives

REBIRTH
OF
NATURE

Audubon Perspectives

REBIRTH OF NATURE

A COMPANION TO THE AUDUBON TELEVISION SPECIALS

Roger L. DiSilvestro

Page Chichester
PRINCIPAL PHOTOGRAPHER

Christopher N. Palmer
EXECUTIVE EDITOR

JOHN WILEY & SONS, INC.

New York   Chichester   Brisbane   Toronto   Singapore

> **T**o those environmental activists—some of whom
> appear in these pages—working locally on the many
> critical issues that affect the quality of all our lives.

Managing Editor: Marcia Samuels
Design: Stanley S. Drate/Folio Graphics Co., Inc.

If you would like to receive information about the National Audubon Society, write to:
National Audubon Society
Membership Department
950 Third Avenue
New York, NY 10022

Library of Congress Cataloging-in-Publication Data
DiSilvestro, Roger L.
   Audubon perspectives : rebirth of nature / Roger L. DiSilvestro ;
Christopher N. Palmer, executive editor ; Page Chichester,
photographer.
     p.   cm.
   Includes bibliographic references and index.
   ISBN 0-471-53208-8
   1. Habitat conservation. 2. Habitat (Ecology) 3. Man—Influence
on nature. I. Palmer, Christopher N. II. National Audubon
Society. III. Title. IV. Title: World at risk.
QH75.D58   1992
333.7'2—dc20   91-36848

Printed in the United States of America

10 9 8 7 6 5 4 3 2 1

FOREWORD

Peter A. A. Berle
PRESIDENT, NATIONAL AUDUBON SOCIETY

The changes that human activities have brought to this planet have made habitat loss the number-one enemy of bird and wildlife survival. And, as Roger DiSilvestro so tellingly states here, if we continue to destroy the natural homes of wildlife, can we realistically believe that we will not, ultimately, destroy our own living place as well?

Protecting habitat, and with it biodiversity, is the most critical issue in environmental conservation today. It is what the National Audubon Society is all about. We began by establishing sanctuaries and hiring game wardens to protect birds and wildlife. We have developed in the past century, so that today we are an activist, chapter-based organization. We are able to act on the community level as well as in national and international arenas on issues of global environmental significance, for example, protecting ancient temperate forests, developing a national energy strategy based on conservation, fighting for wetlands protection, educating people about the impact of population growth on the environment, and protecting endangered species.

Yet sanctuaries, good science, public education, and member involvement at the community level remain the hallmark of Audubon's work.

Working together with dedication, we can win the environmental battles profiled in this book.

PREFACE

Christopher N. Palmer
EXECUTIVE PRODUCER, AUDUBON TELEVISION SPECIALS

The mission of the National Audubon Society's television, books, software, and other mass-media projects is to enlarge and make more active and powerful the global constituency for a healthy and sustainable planet. After people have viewed our programs, we want them to feel inspired to take a more active role on environmental issues. Examples of such actions might include a renewed commitment to recycling, voting with an eye to environmental issues, writing to elected officials to urge them to stronger efforts on behalf of the environment, and joining Audubon or other environmental groups.

What drives this mission? As the twentieth century draws to a close, the human race finds that it has created environmental problems that threaten all life. More active involvement by people to protect the environment is crucial to human and wildlife survival. This involvement must come from a far larger constituency than "card-carrying" environmentalists.

The mission of the National Audubon Society is to protect the wildlife and wildlife habitat upon which our lives depend. Together with more than 600,000 members and an extensive chapter network in the United States and Latin America, our professional staff of scientists, lobbyists, lawyers, policy analysts, educators, and television producers are fighting to save threatened ecosystems and, thus, to restore the quality of life on the planet.

On important issues, from the protection of forests and wetlands to the battle against global warming, we work to influence key decision-makers at all levels of government, from local zoning boards to the United Nations. We manage and protect a nationwide system of wildlife sanctuaries and we promote citizen participation in a host of environmental projects, including community solid-waste management, acid rain monitoring, and local wetlands protection.

Undergirding all this work is our effort to reach out, through television, to the broader public. We produce four Audubon Television Specials a year. They are broadcast both on cable TBS SuperStation (thanks to Ted Turner) and on public television. This book, like its predecessor volume, *Audubon Perspectives: Fight for Survival,* is a companion to eight new Audubon Television Specials:

- "Danger at the Beach," produced by David Clark and hosted by Ted Danson, explores the threats to our coastlines.
- "Wildfire," produced by Tom Lucas and Larry Engel, and hosted by James Woods, investigates the positive side of natural fires.

- "Hope for the Tropics," produced by Pam Hogan and hosted by Lauren Bacall, tells the stories of seven people who are really making a difference in saving the tropical forests of Costa Rica.
- "The New Range Wars," produced by Roger Snodgrass and hosted by Peter Coyote, details the bitter conflict surrounding overgrazing in the West.
- "Great Lakes, Bitter Legacy," produced by Tom Lucas and hosted by James Earl Jones, looks at the longer term impact of toxics in the Great Lakes area.
- "Mysterious Elephants of the Congo," produced by Mark Shwartz and hosted by Jane Fonda, examines the plight of the African forest elephant.
- "Battle for the Great Plains," produced by Judy Hallet and hosted by Jane Fonda, focuses on the Great Plains and asks what the rest of the world can learn from our experience there.
- "Ecotourism," produced by Megan Epler-Wood and George Bell, and hosted by Sam Waterston, explores the role of tourism in saving (or destroying) our wild places.

With this companion book, our goal is to build on the excitement and concern generated by the Audubon Television Specials. This book enlarges on the information contained in the films and tells viewers what they can do to help solve some of our many vital environmental problems. We hope that this book will give readers and television viewers the knowledge and confidence that they need to take action on behalf of sound conservation.

This book was written by Roger DiSilvestro and most of the photos are by Page Chichester. If you enjoy this book, as I am sure you will, you may want to read another book on which they have recently collaborated, *The African Elephant, Twilight in Eden* (also published by John Wiley & Sons). Iain Douglas-Hamilton, the pioneer elephant researcher, described it as the "definitive elephant book—one of the very best."

In getting these television programs and this book to a national audience, the team of people who accomplished this—Audubon, Turner Broadcasting, GE, WETA, PBS, and John Wiley & Sons—are lighting a candle. In 1780 in Hartford, Connecticut, an eclipse one day turned the sky from blue to gray, and the city had darkened over so densely that, in that religious age, people fell on their knees and begged a final blessing before the end came. The Connecticut House of Representatives was in session and many of the members clamored for adjournment. The Speaker of the House, Colonel Davenport, came to his feet and silenced the din with these words: "The day of judgment is either approaching or it is not. If it is not, there is no cause for adjournment; if it is, I choose to be found doing my duty. I move, therefore, that candles be brought to enlighten this hall of democracy."

The television programs in this companion book, along with all the other programs and products that Audubon produces, will help to light a candle so our children can see their way better into the future.

ACKNOWLEDGMENTS

As in the past, thanks must go first to Ted Turner, over whose cable television network (TBS SuperStation) the Audubon programs are aired in addition to their run as a summer series on public television stations nationwide. Turner has remained a staunch ally to Audubon television programming and to the environment, featuring our program on public-land grazing even though faced with organized opposition by well-financed grazing advocates, including serious boycott threats designed to block the broadcast of the program and stifle the free exchange of ideas upon which our society and our liberties are based.

Many of our colleagues at Audubon provided invaluable help in the production of this book, serving both as sources of information and as reviewers for the manuscript. They are Dede Armentrout, Jan Beyea, Dorie Bolze, David Cline, David Henderson, Maureen Hinkle, Ronald Klataske, David Miller, David Newhouse, Ed Pembleton, Carl Safina, Fran Spivy-Weber, and Tensie Whelan.

Two Audubon staff members, Delores Simmons and Ruth Thomas of the Television Department, helped with many of the details involved in travel for this book, making sure that reimbursement checks arrived in the right pockets. They also helped ensure that the many memos, manuscripts, and fax messages involved in producing this book went to and from the right people. Television Department intern Margaret Barker provided indispensable help with the research for the Great Plains chapter. Page Chichester, our principal photographer, also deserves special commendation for doing much of the research for the other chapters.

Any work of this sort involves the time, energy, and interest of many innocent people who fall victim to prolonged interviews, demanding manuscript reviews, long visits to their homes, late-night calls, and last-minute requests. For this and more we give a special thank you to Reynaldo Aguilar, a parataxonomist with INBio in San Jose, Costa Rica; Jim Barborak of the Caribbean Conservation Corporation; Mario Barrenechea, owner of Portico, a Costa Rican door company, and Rodolfo Peralta, a Portico forester; Dr. Pierre Béland and Robert Michaud of the St. Lawrence National Institute of Ecotoxicology in Rimouski, Quebec, Canada; Roger Berkowitz, president of Legal Seafoods, Boston; Manon Bombardier at the McDonald Raptor Center, McGill University, near Montreal, Quebec, Canada; Catherine Bourne, wildlife inspector with the U.S. Fish and Wildlife Service in Baltimore, Maryland; Randy Braun, environmental scientist with the Environmental Protection Agency in Edison, New Jersey; information officers James Caldwell, Ollie Van Crump, Diane Drobka, Wendell Peacock, and others who worked the 1990 fire at Yosemite National Park; Tanya D'Ambrosio, who, as an official of the Costa Rican Board of Tourism, made the author's travel in that nation a rare delight; Don Despain, research biologist at Yellowstone National Park; John Dinga and Mike Borek of Harbor Explorations at the University of Massachusetts at Boston; Steve Dobrott, wildlife biologist at Buenos Aires National Wildlife Refuge in Arizona; Richard Donovan, a World Wildlife Fund forestry consultant; Iain Douglas-Hamilton, an elephant researcher in Kenya; Jim Fish, a founder of the Public Lands Action Network in Albuquerque, New Mexico; Scott Franklin and Donald Pierpont, fire captains for the County of Los Angeles; Lynn Gooding, inspector, Washington State Department of Ecology in Olympia, Washington; Jerry Holechek, professor of range sciences at New Mexico State University, Las Cruces; Jerry Hoogerwerf, a pilot with Project Light Hawk, who provided a photographic flight over Rio Puerco, New Mexico; David Hull, water-resource specialist in Arcata, California; Steve Johnson, founder of Native Ecosystems, an environmental consulting service in Tucson, Arizona; Dianne Jurgen, a public participation assistant with the Massachusetts Water Resources Authority in Boston; Mike Kennedy, of the Sewer Utility Division of the City of Tacoma, Washington; Rodrigo Gamez Lobo, director of INBio in San José, Costa Rica; Gilbert Lusk, superintendant of Glacier National Park, Montana; Stephen Pyne, a fire expert and history professor at Arizona State University-West; Peggy Rice, industrial-waste specialist with the Tacoma, Washington, Metro Environmental Laboratory; Bob Schoelkopf, head of the Marine Mammal Stranding Center in Brigantine, New Jersey; Rick Stephanic, fire information officer, and others who worked the 1990 Custer National Forest fire in Montana; Mike Stoner, of the Environmental Protection Agency in Seattle, Washington; Alvaro Ugalde, superior director of the Department of Wildlands and Wildlife in the Costa Rican Ministry of Natural Resources, Energy, and Mines; Alvaro Umana, vice-president of the Center for Environmental Study in San José, Costa Rica; intrepid iguana researcher and popularizer Dagmar Werner; Chris Wille and Diane Jukofsky, founders of Informacion Tropical, a rainforest news network, for their patient and endless help while the author was a guest in their home in San José, Costa Rica, and long afterward—perhaps now the requests for further data will end; Jim Winder, a rancher in Nutt, New Mexico; and Ted Wood, photographer and tour guide extraordinaire in Jackson Hole, Wyoming.

Heartfelt thanks for their singular professional skills and profound patience with an often sagging deadline schedule go to David Sobel, the editor for John Wiley & Sons, and Marcia Samuels, the production whiz who navigated the photographs and manuscript through various critical production channels at Wiley. The very best in promotional support came from Wiley's Peter Clifton and Los Angeles public relations consultant Caroline O'Connell, both of whom have been invaluable to the success enjoyed by some of the author's previous books.

The narrators of the Audubon specials included in this book gave generously of their time from busy schedules and demanding professional lives. They are Lauren Bacall, Peter Coyote, Ted Danson, Jane Fonda (whom we thank twice, for hosting two programs), James Earl Jones, Sam Waterston, and James Woods.

No book could ever be completed without the diligent support of those on the home front. The support troops during the most strenuous battles of the book war were Jeanne Marquis, Ini Chichester, and Gail Shearer, the significant others of the author, photographer, and executive editor respectively.

And thanks to all our readers and viewers, without whose continued interest and support neither the television nor book projects would be possible.

CHRISTOPHER N. PALMER
ROGER L. DiSILVESTRO
PAGE CHICHESTER

CONTENTS

Audubon Perspectives

REBIRTH OF NATURE

LOOK HOMEWARD, CONSERVA-TIONIST

I n this book we are going to talk a lot about habitat. At first blush that may seem a dull subject, the sort that puts college science students into deep and restful slumber. But the subject is far from dull. Indeed, it lies at the heart of today's most critical wildlife conservation issues.

Human activities threaten to destroy up to half the world's species within the next few decades, perhaps as many as 15 million different types of plants and animals. Most of these species are being wiped out not by overhunting or poaching but by loss of habitat. When habitat disappears, species disappear. This destructive loss of habitat is happening even though knowledge accumulated about the biological workings of the planet has shown clearly how intimately interrelated are all species, how interdependent they are for survival—how interdependent *we* are for survival. But habitat destruction goes on apace, as if humankind were either compulsively suicidal or hopelessly reckless.

A *habitat*, in the very simplest of terms, is a place in which an organism lives. It provides the organism with all its biological needs. These needs vary from species to species, but usually include food, water, atmosphere, shelter, and enough space to allow the individual organism to take advantage of those essentials without competing too strenuously with other members of the same species.

The term *habitat* covers a gamut of living places. A forest is a habitat, home to a tremendous variety of creatures—trees, shrubs, grasses, fungi, bacteria, protozoa, lichens, birds, insects, mammals, reptiles, fish, amphibians—hundreds, indeed thousands, of species. A puddle of water is a habitat, home perhaps to insect larvae, tadpoles, microscopic plants and animals, a whole array of living things. The ocean is a habitat, the desert is a habitat, the human intestine is a habitat. Taken to its extreme, the term *habitat* almost seems to encompass everything you can lay your eyes on and many things you cannot.

It probably could go without saying that humans are as habitat dependent as any other living thing. We need food, water, shelter, and the rest as much as does any plant or nonhuman animal. Our specific biological needs make it impossible for us to live in regions deficient in one or more essential. Think of living in the heart of the Sahara or the Antarctic or cast adrift in the ocean. These places, without a lot of hardship and hard work, are not for us. We have been adapted through the long course of evolution to certain conditions of habitat. We know this implicitly. When we look at other planets, with scorching surfaces or atmospheres of methane gas, we know that we could not live there, nor could anything else on Earth. We say, "Life as we know it cannot exist there." Life as we know it can live only on Earth, because Earth, or some portion of it, is our habitat. The link between life and planet is, to all practical purposes, unbreakable. Nevertheless, we have done a poor job of protecting the habitats upon which we depend.

◄ N ew York City late-night traffic.

Mushrooms growing in a Caribbean island rainforest can be a habitat for uncounted insects and microorganisms. Rainforests cover only a small percentage of Earth's surface, yet host nearly half the world's species.

The Mutuality Factor

Species are linked not only to their habitats, but to one another. This happens on a large scale, as when a species such as the koala becomes adapted to eating only the leaves of the eucalyptus tree or the black-footed ferret to eating primarily one species of rodent, the prairie dog. But vital bonds between species also exist on a much more subtle level.

Some classic examples are found in the tropical rainforests. Because air movement in rainforests is limited, many tropical plant species depend upon animals for pollination and so have developed a number of mechanisms to enhance the process. One case is the durian, a tree that produces fruits sought by a wide variety of species, including tortoises, elephants, deer, rhinos, bears, and even tigers. About 45 percent of the durians that flower in Malaysia are pollinated by bats. Not surprisingly, the durians have become closely adapted to the bat's daily schedule: The durian's stamens are receptive to pollen only at about 8 P.M., when the bats are active. This relationship shows how the survival of various rainforest species are intimately linked. Without the bats about half the durian species would vanish, and the animals that feed on its fruit would lose an important food. This makes many animal species, as well as the durian, dependent on the bat. And since the bats feed on many other species when the durians are not flowering, the bat's survival depends on the survival of a large number of other plant species. Thus the durian's fate is irrevocably tied to that of a myriad of other rainforest plants, as is the fate of several species of forest animals.

One of the most fascinating relationships between plant and animal concerns pollination in fig trees, which number some 450 species in Malaysia and Indonesia alone. Figs produce three types of flowers—male, female, and gall flowers. The gall flowers are shaped like flasks and bloom inside a larger flask-shaped organ, called a pseudo-fruit because it resembles the fig's seed-bearing fruits. Tiny wasps live inside the gall flowers, where they were deposited as eggs. When the wasps emerge, the wingless males mate with the females and then die. The females then leave the pseudo-fruit, coming into contact with pollen-producing male flowers as they go. The production of the pollen coincides with the emergence of the wasps, which exit the pseudo-fruit covered with pollen. The pollen-laden females fly to other fig trees and, preparing to lay eggs, enter the flower-bearing pseudo-fruits, where their contact with female flowers results in cross-pollination.

Another classic case of intimacy between insect and pollinating plant is that of the bee and the orchid. The elaborate flowers produced by orchids are mechanisms for attracting insects and for covering them with pollen.

A prime example begins with the courtship ritual of Central American euglossine bees. Male euglossines attract females by performing a courtship flight in shafts of sunlight. But it is not the bees' dazzling color alone that attracts the females. The males also wear a fragrant, oily perfume obtained from an orchid that produces it to attract the male bees. The bees land on the flower to collect the oil and, at the

Coastal development near Atlantic City, New Jersey, spells doom and destruction for wetlands and wildlife. Such shortsighted activity often is a self-defeating proposal as well, since many houses are built on eroding sands.

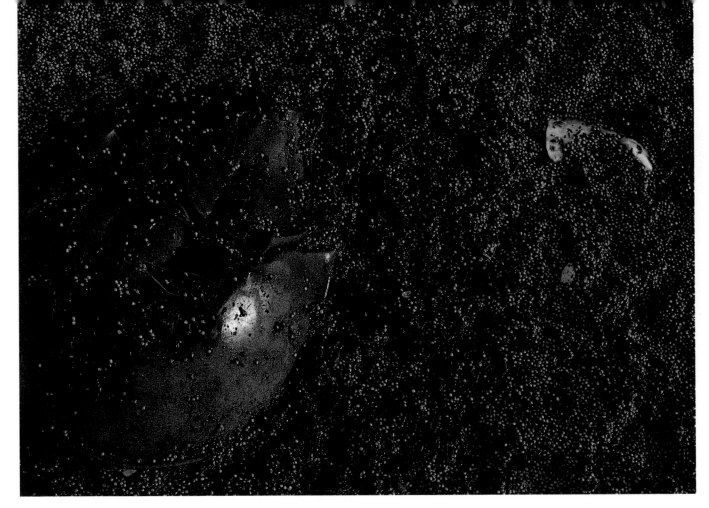

same time, receive a load of pollen. As they fly from flower to flower in search of oil, they spread the pollen about.

What happens in the flower as the bee tries to collect the oil can be quite complicated. Good examples occur among the 20 or so species of Central American bucket orchids. A male euglossine bee that lands on the edge of the orchid's bucket-shaped flower finds it cannot get a firm foothold on the waxy surface. The bee slides into the bucket, which is filled with a fluid that drips into it every day at dawn from a gland located on the bucket rim. The bee struggles to get out but finds only one avenue of escape, a little step that allows it to crawl out of the fluid. The step leads into a tunnel, which leads to the outside. Willy-nilly, the bee scuttles down the tunnel. As it emerges from the tunnel into sunlight, another organ snaps down on it, locking it in place. At this moment, a packet of pollen is glued to its back. When the bee is finally released, the pollen stays in place until the bee enters another flower of the same orchid species. It then goes through the same process, but this time, as it moves down the escape tunnel, a hook in the roof snatches away the pollen and the flower is fertilized. As for the bee, it receives another pollen packet upon exiting.

In many such cases, orchids and bees have co-evolved so that the bee has certain anatomical features that adapt it to pollinating only one or a few orchid species, and the orchids have adapted so that they can be pollinated by only one or a few bee species. This selectivity, and the elaborate mechanisms that some orchids use to apply pollen to bees, help ensure that each orchid flower receives pollen only from others of its own species. This prevents infertile crossbreeds. It also indicates that there is tremendous evolutionary pressure for individuals of widely scattered species to evolve means by which to find one another.

Bees are also crucial to the prosperity of the Brazil nut tree. In this case, the bees are allied with agoutis, a type of large rodent, and with an important commerical industry: About $20 million worth of the nuts are exported to the United States alone each year. The nuts are harvested from wild trees because attempts to create Brazil-nut plantations have failed. Apparently the trees can grow only in mixed forests that contain a wide array of species. Among these species are euglossine bees, the same ones that pollinate orchids. How Brazil-nut trees are pollinated is still a mystery, but the bees may play a role in successful Brazil-nut reproduction, since it is well known that the trees will not reproduce in areas that lack euglossine bees. This also means that the trees will not grow in areas that lack the other plants, such as orchids, upon which the bees feed. Moreover, Brazil nuts cannot germinate without the help of the agoutis, the only animals known to crack open the hard-shelled pods that contain the Brazil nuts. Reproduction of Brazil-nut trees therefore also depends on the presence of all the plants and animals that the agouti needs to survive. The rainforest is much like a house of cards. The whole may begin to collapse if a single critical part is removed.

Virtually every habitat type has examples of crucial bonds between species. In the deserts of the U.S. Southwest, the decline of the Mexican free-tailed bat, caused by human disturbance of bat caves, may be jeopardizing survival of the giant saguaro cactus, which depends upon

Ants tending planthoppers in Chilamante, Costa Rica, provide an illustration of co-evolution. Species that share a habitat develop adaptations to one another, increasing their chances for survival.

The moon races ahead of dusk at Saguaro National Monument near Tucson, Arizona. Disappearance of the Mexican free-tailed bat, which pollinates the tall cacti, could mean the end of the saguaro.

the bat for pollination. Loss of the cactus could in turn jeopardize bird species that nest within it. As human inroads drive species to extinction, habitats and species complexes are undermined. The undermining of habitats in turn causes the loss of more species, which further undermines habitat, and so on. Even a cursory look at the world's grasslands, forests, wetlands, and seas shows that humanity, in its sweep across the planet, is ruining vast natural systems.

Less apparent, however, is humanity's dependence on the natural environment. As we literally cut the ground out from under millions of species, we are eroding our own turf, too. This may not seem apparent when we are outfitted in our Burberry's suits, ensconced in dimly lit restaurants with starched linen tablecloths, the taste of dry white wine cleansing our palates of the cream sauce of the lobster Newburg. It is easy to forget that the wine was pressed from sunlight-nurtured grapes that sprouted from a certain type of soil, that the lobster came from an ocean floor where the Darwinian struggle for survival was not quite the abstraction that urban living makes it seem. We are still linked to powerful natural forces whose destruction we cannot ignore with impunity.

The Human Factor

Though the human species is dependent for its survival on healthy habitat—both globally and locally—we are also perhaps the most adaptable animal that ever existed. We seem able to meet successfully almost any challenge the natural world poses. This is because our adaptations to environmental changes, unlike the adaptations of most species, are not restricted to genetic mutation, which is a slow process. Humans can adapt behaviorally. We do not need to evolve warm coats of fur if temperatures decline; we simply put on more clothing. If our eyes focus so poorly that we can hardly see, we need not fail in the evolutionary struggle; we merely put on glasses.

Our extreme adaptability allows us to invade all types of habitats, from hot deserts to polar ice. We may have entered the evolutionary world as creatures of tropical savannahs, but we are now global animals. Probably no other terrestrial species has ever ranged so widely.

Our behaviorial adaptability has allowed us to burst the bonds of natural restrictions on population growth and habitat degradation. In virtually all other species, ecological limiting factors, such as the availability of water or food, will keep populations from overrunning their habitat. But in human industrial societies, when some necessary resource is scarce, we simply import it from another habitat. The arid southwestern United States could not support its millions of people were water not taken from underground aquifers and from distant rivers and lakes. Los Angeles and its suburbs have survived by draining water from neighboring states. Intensive agriculture has let us escape the limits placed upon population growth by the restricted availability of wild food sources. Our behaviorial adaptability has increasingly insulated us from the natural controls that keep species from destroying their habitat by sheer numbers alone. Many of us even believe that we are free of natural constraints, that we need no longer worry about the condition of the world beyond the human doorstep, that technology will fix whatever problems it creates.

However, the stay of execution that we have received from the impact of natural controls is at best temporary and at worst illusory. Humanity is still dependent upon its habitat, regardless of how varied, for all biological needs. As pointed out in the Worldwatch Institute's environmental overview *The State of the World,* three biological systems sustain us—croplands, forests, and grasslands. They supply all raw materials for industry, with the exception of fossil fuels, and all foods except those that we take from the sea.

The importance of croplands, forests, and grasslands to human society is based on photosynthesis, the ability of plants to produce edible or durable materials from sunlight. An estimated 40 percent of Earth's photosynthetic products now go to our consumption or have been lost because of environmental degradation we have caused. As we have commandeered a larger share of the world's produce and turned vast portions of natural ecosystems into either agricultural fields or waste lands, we have wiped out the natural resources that provided the essentials of habitat to many, many species.

We can see the process occurring everywhere. Grasslands on every continent are being destroyed by overgrazing. In 1950, Africa's human population of 238 million had 272 million cattle grazing on African grasslands. A quarter century later, the human population was 600 million and grasslands were being taxed with 540 million cattle. Overgrazing—the process of putting more livestock on the land than the land can sustain while renewing itself in good health—is turning grasslands throughout the world into deserts. In India, the need for cattle feed will reach an estimated 700 million tons by the end of the century, but India can produce only 500 million tons. More than half of U.S. grasslands managed by the federal Bureau of Land Management are in fair to poor condition because of grazing. Roughly 99 percent of the tall-grass prairies that once cloaked the midwestern U.S. and Canada have been replaced with crops, which cannot support the variety of wildlife species found in natural grasslands.

Forests have been appropriated for our use, too, with little thought of tomorrow. Clear-cutting and burning are rapidly wiping out the tropical rainforests. Air pollution, particularly acid precipitation stemming from human activities such as the burning of fossil fuels and

A wary zebra takes flight in Amboseli National Park, Kenya. The East African nation's savannahs are threatened by human overpopulation and cattle grazing. Cows compete with wildlife for grasslands outside the protected parks.

▶

A bison surveys his domain in the National Bison Range, Montana. A large bull can grow to six feet at the shoulder, weigh 2,000 pounds, and run as fast as 32 miles an hour. His is a temperate grasslands habitat, prone to cold winters, hot summers, and steady wind.

wood, is threatening vast forest tracts in the United States, Canada, and Europe. An inventory by the U.S. Forest Service found that dead pines in southeastern U.S. forests increased from 9 percent of all trees to 15 percent from 1975 to 1985.

About half the wetlands in the United States at the time of European discovery have been drained, mostly for agriculture. This takes wetlands away from creatures that need marshes and swamps. Nearly half a million acres more are destroyed each year. Those that remain are often poisoned with human-generated pollutants. Even the ocean is showing signs of toxic stress. Beaches in populated areas are increasingly awash with garbage and debris that was dumped offshore. The disposal of toxic chemicals in the oceans is also coming back to haunt our shores. Many marine biologists believe the loss of an estimated half of the coastal bottlenose dolphin population along the U.S. Eastern Seaboard in 1987 was related to heavy pollution.

But more is at stake than healthy trees and clean beaches. As plant and animal species vanish because of habitat destruction, human society loses unimagined potential resources for new crops, industrial products, domestic goods, and medicines. Also, as we lose pristine wildernesses and beautiful vistas, we lose a certain amount of aesthetic well-being. We lose the option of being able to go to places where we can escape the inroads that humanity makes upon humans.

A Matter of Degree

There are those, of course, who would argue that we need oil and gas more than we need pristine coastlines. And there are those who, never venturing beyond the smell of urban air, may be hard-pressed to see the need for open spaces. They see the loss of a forest in the

A snowy egret wades the wetlands for its minnowy meal at Prime Hook National Wildlife Refuge, Delaware. Wetlands and other islands of life are disappearing at an alarming rate because of development and agriculture.

Northwest or a prairie wetland in Minnesota as insignificant. And if we were indeed losing only a marshland here and a portion of forest there, the losses would not threaten the integrity of the globe or human survival. But the destruction does not concern a few isolated spots here and there. We are threatening the full gamut of habitats that sustain our very lives. Polar regions are degraded by oil development. Coastal areas all over the world are being lost to development and pollution. There is probably not a river or stream in the industrial world that has not been degraded to some extent by pollutants. Our indiscriminate consumption is turning grasslands and forests into deserts. Underground aquifers, from which many U.S. communities draw water supplies, show increasing amounts of agricultural pollutants. The production of wood is declining in U.S. and European forests that have already been cut twice—apparently the trees simply are not growing as fast as they used to. Not a single habitat type is unaffected.

It seems unlikely that we can escape the impact of this great range of environmental destruction. Eventually, such a significant portion of our global habitat will be degraded that the planet will lose its ability to support the large numbers of people dependent on it. To maintain the current standard of living, populations will have to shrink. This does not seem likely to happen in most parts of the globe. It is more likely that prices for increasingly scarce foods will skyrocket, that the poor will get poorer. We cannot survive at an optimum level when our habitat declines, when we can no longer get from it all that we need in the way of food, water, and living space.

This book surveys a number of habitat types that are suffering beneath the heel of human rule. It outlines the many threats that jeopardize the integrity of human society and the health of the Earth. Fortunately, the story is not one of unrelenting gloom. As knowledge accumulates about the workings of the natural world, growing numbers of informed people are questioning the traditional, exploitative approach that is consuming the world's resources.

The next decade will be a vibrantly exciting one, as any struggle for survival is exciting, and at the center of it will be those who are working to protect the globe that gave and continues to give us life. This book—as well as the television programs upon which each chapter is based—is both a report on the activities of these dedicated souls and a salute to them and to the hope and encouragement they give to others.

A shrimp boat plies the waters off the coast of South Carolina at dawn. Such areas, from the low-water line to 650-foot depth, are the ocean's seafood factories, where plentiful nutrients and light are available to ocean-going species. ▶

A sign offers a poignant warning and a symbol for the futility of attempting to tame the world. If people continue to ignore Nature's warnings, we stand to be buried for our insolence.

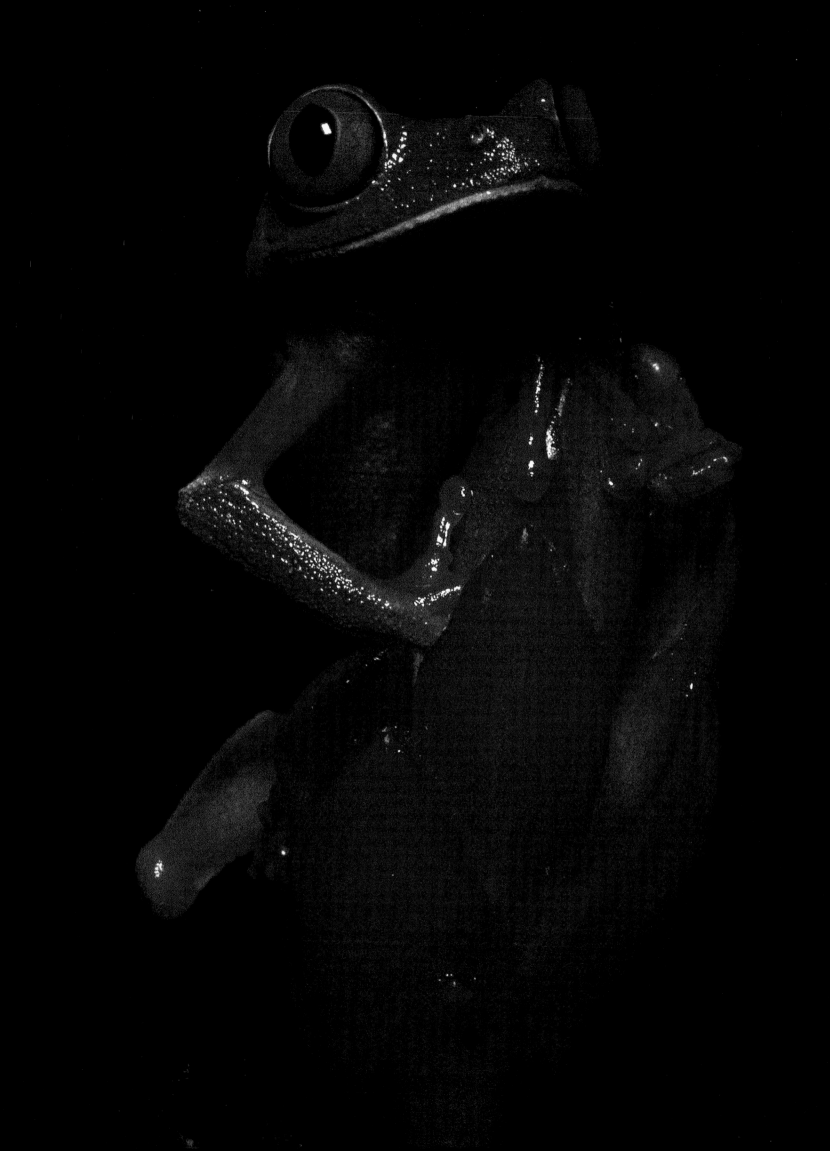

CHALLENGE AND CHANGE IN COSTA RICA

ost books, magazine articles, lectures, and television programs about tropical rainforests attempt to impress upon us the tremendous diversity of life found there. The litany of superlatives is doubtless familiar to anyone interested in the rainforests and their fate:

- Rainforests cover less than 10 percent of the Earth's surface, but house roughly half the world's 5 million to 30 million species.

- On a single mountain in Borneo grow five times as many oak species as in all of Europe.

- Colombia—a third the size of the United States—has 25 percent more plant species.

- From a quarter acre of Colombia's Chocó rainforest spring some 200 plant species, about eight times as many as grow in all of New Hampshire.

- A quarter acre of Latin American rainforest may bear 10 to 20 times as many species as an entire North American forest.

- A single park in Costa Rica shelters more bird species than all of North America.

- Costa Rica's 1,800-acre La Selva Forest Reserve contains twice as many tree, bird, mammal, reptile, amphibian, fish, and insect species as California, which encompasses more than a million acres.

But none of these statistics prepare a visitor for the great tumult of living things that awaits in the rainforest. From the ragged green tops of the highest trees to the damp-rooted ferns of the forest floor, the deeply shadowed rainforest is a rushing cascade of plants and animals, a heaping swarm of living things, a veritable avalanche of life.

Even a small tract of rainforest, such as Costa Rica's 1,200-acre Carara Biological Reserve, vibrates constantly with the movement of living things. Carara lies along a highway that winds through rolling hills that overlook the black-sand beaches of Costa Rica's Pacific shores. The surrounding hills have been denuded of their deep forest and turned into farms and cow pastures, but at Carara the old order still survives. Wander into the verdant forest, leave the highway just 50 yards behind, and you enter another world. A different bird species appears every few seconds—hummingbirds, scarlet macaws, toucans glossy black with long, rainbow-hued bills. A coatimundi, tropical relative of the raccoon, wanders down out of the trees to forage at leisure near visitors, showing neither fright nor concern. A troop of white-fronted capuchins—the monkeys preferred by old-time organ grinders—pauses in its treetop rambles to look at the larger primates below, who gaze back through binoculars. An emerald-green and black frog scurries through the leaf litter, slim and smooth and beautiful as jade, but deadly. The mucus exuded from its skin has been used for centu-

◀ A red-eyed tree frog clings to its perch in Costa Rica.

ries by native rainforest peoples to poison the tips of arrows. Butterflies with wings the size of a man's hand pirouette over the trail, land on the ground to feed on fallen fruit. A black-billed cuckoo, a species familiar to North Americans, races through the trees, living evidence that the forests of Central America are the home of northern species, too, and just as important to their survival as are the woodlands and plains of Canada and the United States. Every leaf seems crowded with grasshoppers, grasshoppers more beautiful than a North American visitor might ever expect a grasshopper to be—metallic blues, emerald green trimmed with saffron and scarlet.

Trees drip with life. Many branches are heavy with bromeliads, basketlike plants that draw sustenance from animal and vegetable matter that falls into them. Vines entwine tree trunks, wrapping so tightly that the trunks themselves are barely visible. Split-leaved philodendrons crawl up the trunks, philodendrons so full of the explosive power of life that each leaf is many times the size of a leaf on the houseplant variety. Mosses and lichens crowd the vines for living space. Every nook and cranny is home to some plant or animal. Tap the trunk of a cecropia tree and from every crack, crevice, and seam will swarm the warriors of an ant colony, ready to do battle with anything bold enough to invade the tree in which they live.

Carara is a kaleidoscope of living things. It offers sights and sounds in such quick succession that your senses cannot hope to perceive them all. It is as if the forest were some masterpiece of evolution, some demonstration of the true genius of nature and adaptation. Its complexity is so vast and deep that our best scientists have barely begun to fully comprehend it. To know the rainforest is not to understand it, but to be awed by it. It humbles by comparison any work of humankind.

And now it is vanishing. Carara is a small reserve, beset by poachers and land-hungry farmers and ranchers. Its boundaries are ignored by axe-wielders and riflemen, and the Costa Rican government cannot afford the protection the reserve needs. Everything that you encounter there—from the brilliant scarlet macaws flying raucously overhead to the sprightly monkeys to the tiny gladiatorial ants lying in wait within the cecropias—is at the moment a mere phantom shimmering before your eyes, doomed to disappear. Carara, says Alvaro Ugalde, head of Costa Rica's Department of Wildlife and Wildlands, is too small to survive intact the pressures put on it by poachers and agriculturalists.

The plight of the Carara Biological Reserve is the plight of rainforests all across the globe. Every year, worldwide, an area roughly 20,000 times the size of Carara is swept away by axe, saw, and fire. At that rate most of the globe's rainforests will be gone within the next 30 years.

One hope for stemming this destructive tide lies in Costa Rica. The sources of that hope are the many new protective forest projects that the Ticos—as Costa Ricans call themselves—are initiating. Most of the projects are experimental, but if they prove successful they may become models that other tropical nations can use to begin sound protection and management of their own rainforests.

▶

An iridescent flash reveals a blue morpho butterfly among ferns on the forest floor in Tanto Domingo, Ecuador. Rainforest destruction worldwide threatens to wipe out remaining forests within the next 30 years.

Why Save Rainforests?

It is no exaggeration to say that the world's tropical rainforests touch the lives of virtually every resident of the globe every day. In the tremendous diversity of living things that the forests produce lie vast benefits for human society.

The rainforests are a natural treasurehouse where diesel fuel flows from the trunks of trees and miracle drugs lie hidden in the leaves of obscure plants. Scientists are constantly searching the forests for these riches, some of which already have been put to use. Rainforests supply ingredients for birth control pills, tranquilizers, heart medicine, laxatives, and stimulants, and provide drugs used to treat hypertension, leukemia, menstrual problems, mental disorders, syphilis, skin disease, rheumatoid arthritis, sciatica, acute bronchitis, glaucoma, seasickness, and more. But this is a sparse beginning. Science has closely tested the medical potential of barely 1 percent of the rainforests' plant species. Perhaps thousands of other medical applications still lie undiscovered, at least by modern science, in those places so long thought of as jungle wastelands.

The number of plants that forest peoples use as curatives is a sign of the tremendous medical potential hidden in the rainforests. Natives of the Amazon rainforest use no less than 1,300 plant species for curing various illnesses. Herbalists in Southeast Asia use five times that many—some 6,500 species—in their treatments. In India 1,000 wild species are used in medical treatment, in Thailand 500, in the Malay Peninsula more than 2,400. Some of the folk treatments may be grounded in superstition and completely worthless. But if even a small fraction of these plants are medically useful, they will be a great bounty for human society. It is increasingly clear that the forests are like huge medicine chests whose doors have hardly been opened. A single survey of 1,500 plants from the Costa Rican rainforest revealed that some 225 of them could potentially produce anticancer drugs. Researchers believe that as much as 10 percent of the 90,000 plant species thought to grow in Latin America will yield anticancer drugs.

Tropical rainforests provide not only cures but sustenance.

TROPICAL RAINFOREST

THE ISSUE Rainforests throughout the world are being destroyed so rapidly that if current losses continue, most forest will be gone by 2025.

THE CAUSE The forests are being cut for lumber and burned to make way for crops.

EFFECTS ON WILDLIFE Millions of species will be lost as the forest disappears.

EFFECTS ON HUMANS Those who dwell in rainforests will lose the context of their cultures as their habitat disappears. Farmers who cut the forests for croplands will ultimately face economic ruin because most rainforest soils cannot sustain agriculture for more than two years. Denuded, exhausted farmlands are often used for livestock pasture, but this use too is limited to only a few years. As forests are lost, humanity will lose many potential food crops and medicinal products. We will also suffer a terrific aesthetic loss.

EFFECTS ON HABITAT When tropical rainforests are destroyed, the loss is probably permanent. The heavy seeds of many plants and their intolerance to bright light make it difficult for forests to regrow in denuded areas. Complete renewal, even if possible, might take thousands of years. Loss of wildlife species that play a role in plant reproduction—for example by dispersing seeds or pollen—may make it impossible for many surviving plant species to reproduce.

A "hot lips" flower offers the forest visitor an explosion of color in Braulio Carrillo National Park, Costa Rica. Rainforest plants provide ingredients for many medicines and consumer products, but science has only begun to scratch the surface.

They have pleased our palates with a myriad of foods, among them coffee, cocoa, avocados, rice, millet, peanuts, yams, pineapples, bananas, mangos, papayas, cassavas, lemons, eggplants, cashews, oranges, and tea. Many, many more foods lie virtually untouched in the forest larder. We make common use of perhaps 50 fruit species, but the rainforests offer another 2,450 edible fruits. Among them are the Malaysian mangosteen, often touted as the best-tasting fruit in the world; the Southeast Asian pummelo, the largest known citrus fruit, bigger than a grapefruit; and the South American soursop, which weighs up to 8 pounds.

New sweeteners, too, are coming out of the rainforests, sweeteners made of proteins, which makes them considerably more desirable than the sugar-based sweeteners that plague human health in so many ways. At least three plants from West African forests produce protein sweeteners that are hundreds of times sweeter than sucrose and even saccharin. The Japanese are already using derivatives of wild West African and Latin American rainforest plants to sweeten jellies, chewing gum, beverages, soups, and even meats and fish. One of these sweeteners is calorie free, making it a good addition to foods presently sweetened with calorie-laden sugars. It comes from a Paraguayan plant whose crushed leaves have long been used by forest natives to sweeten their meals.

Wild tropical plants offer not only food for people but also renewed vigor for domestic crops. Modern agriculture as practiced in Europe and North America is based on a scant handful of crops, perhaps two dozen species. Fully two-thirds of the world grain crop comes from only three species—corn, wheat, and rice. These are the plants that our Neolithic ancestors first farmed some 10,000 years ago. Over the centuries, these plants have been refined by selective breeding to produce much larger edible parts than their wild ancestors ever did. However, selective breeding reduced their genetic variability. Domestic plants must be periodically infused with fresh genes to maintain productivity and to increase resistance to disease. Rice crops, for instance, have been protected from grassy stunt virus by genes taken from a species of wild rice. Not a single one of some 10,000 varieties of domestic rice would have genetic resistance to the disease were it not for the genes from the wild strain. Wild genes can also improve the appeal of agricultural products. A naturally caffeine-free variety of wild coffee bean is being used to create caffeine-free domestic beans. Similarly, the taste of cocoa is being improved by infusions of wild genes.

The benefits offered by the rainforests can be measured in real goods right now. A partial list of the many products that originate in tropical forests shows how profoundly the forests have influenced our lives. The list includes shampoos, sun screens, lubricants, clothing fabrics, flavorings, polishes, steroids, pesticides, edible oils, dyes, plastics, solvents, cosmetics, colognes, antiperspirants, perfumes, ice creams, condiments, detergents, paints, varnishes, and a wide variety of others goods, including natural rubber. Many tropical plants produce natural repellents and pesticides that do not pose the toxic threats to human health inherent in most chemical pesticides. These natural compounds could be used in place of the chemically engineered pesticides to fend off the insect pests that consume roughly half of the world's crops. Some tropical rainforest plants can be processed to produce alcohols and oils that are suitable substitutes for fossil fuels, such as petroleum and coal. During World War II, the Japanese pressed the nuts of a Philippine tree, appropriately called the petroleum nut tree, to produce an oil that was used as fuel in tanks. One tree in the Amazon produces a sap so similar to diesel fuel that it can power an engine without any processing. A search for fuel-generating plants could turn up a thousand such species in Latin America alone. They could be used to wean the world from environmentally damaging fossil fuels and to bolster the economies of developing nations presently committed to importing large quantities of oil.

The rainforests of the world are certainly among the most rewarding of all the natural resources that humankind has ever had at its disposal. If properly protected, husbanded, and cared for, the forests could provide human society with as yet unimagined benefits.

What Is a Tropical Rainforest?

Tropical rainforests are a product of climate. They grow only in the warm, wet parts of the world, where the average annual temperature is about 80 degrees Fahrenheit and where at least 72 inches of rainfall drenches the earth every year. Rain is so important to the rainforest that the forest cannot survive long without it. Rain must fall year-round, amounting to no less than three inches in any given month.

The need for warmth and water restricts tropical rainforests almost entirely to a global belt that lies between the tropics of Cancer and Capricorn. Outside that region, mean temperatures drop to about 68 degrees Fahrenheit during the coolest months, too chill for tropical rainforests.

Altitude, too, affects climate and renders some tropical regions off-limits to rainforests. At roughly 3,000 feet above sea level, mean low temperatures in the tropics fall below 68 degrees. This height marks the border at which rainforests yield to montane forests.

The rainforests also are a product of their soil. The rainforest's tremendous diversity of species suggests that the forests grow on rich soils. Early explorers believed that if the forests were felled, the tropics could be turned into rich farmlands. Modern science has shown decisively that in most of the tropics nothing could be further from the truth.

Soil is created by the breakdown, through millions of years, of underlying rock, and the quality of a soil depends in part on the material from which it was first produced. The soils in large parts of the tropics originated from rock high in feldspar. As this type of parent rock weathers and disintegrates, rainfall drains through it, removing elements such as aluminum hydroxides and silicates but leaving iron oxides, quartz, and elemental aluminum. This process ultimately produces a type of soil called *laterite*. Laterites are sticky and reddish

A flower lands on the decayed litter of the rainforest floor, in Costa Rica's LaSelva Biological Reserve. High rainfall in the forests leaches nutrients from the soil, leaving a poor basis for agriculture.

when wet and are therefore called *tropical red earths.* Some of them are hard as rock when dry.

Laterite soils characterize about 66 percent of the soils in tropical zones, all of them lacking in nutrients. Another characteristic soil is an acidic, sandy type, also poor in fertility. The constant, year-round rainfall is one major reason for this lack of nutrients. Rain water dissolves plant nutrients that lie at the earth's surface and around plant roots. As the water percolates downward, the nutrients are carried away. The flow of nutrients away from the roots contrasts sharply with nutrient flows observed in the more distinctly seasonal temperate zones. There, during dry periods, evaporation from the soil surface draws water upward, bringing nutrients to the upper layers and leaving them near the roots as the water evaporates.

By any measure, fertile soils are scarce in the tropics. Only 13 percent of tropical soils in Latin America and 24 percent in Africa can be called fertile. Asia does somewhat better: Some 40 percent are fertile. However, almost all of these areas have already been exploited for human use.

Rainforests thrive not because of their soil but in spite of it. The poor soils require most tropical plant life to draw nutrients primarily from decaying organic matter—decomposing trees and dead animals, rotting fruit, fallen branches and leaves. The rainforests feed upon their own dead. Millions of rainforest creatures help the process of decay. Hundreds of species of termites are constantly at work, breaking down dead wood. Ants, worms, millipedes, beetle larvae, and thousands of other species reduce dead plants and animals into tiny fragments and provide fecal matter as well. Leaf litter is devoured by worms and other invertebrates of the forest floor. They digest it and produce dark, sticky humus, which cements soil particles together, increasing the soil's ability to hold water and to absorb minerals. Eventually the process of decomposition reduces dead things to their chemical components, which plant roots can readily absorb.

The confinement of scarce nutrients primarily to the upper layers of the soil causes the roots of tropical plants to sprawl shallowly near the surface. Rainforest trees are so starved for nutrients that many of them have evolved an unusual means for tapping food sources outside the soil—they grow roots from their limbs. These crown roots tap into organic debris that accumulates on lower branches, such as the decomposing materials that build up among the roots of epiphytes, plants that live on the branches and trunks of rainforest trees. In some trees, crown roots actually grow into the tree's own trunk and absorb nutrients stored there. This is particularly productive in trees with hollow trunks, because animals that shelter in the trunks usually leave behind a variety of nutritious debris. By rooting into its own trunk, a tree can avail itself of these nutrients. Some trees have even evolved a mechanism by which the core of the trunk normally rots away as a means for releasing nutrients locked up there.

Human involvement in rainforest ecology often cuts short the process of nutrient recycling. If a tropical rainforest is clear-cut and the fallen plants removed, the nutrients the plants contain are never returned to the soil. The soil quickly becomes depleted, and the forest becomes an ecological wasteland. If the plants are burned in place, their nutrients are released in one great flood. This provides momentary benefits, but in the long term it is just the beginning of another ecological disaster. Rainfall quickly leaches the nutrients from the charred soil, yielding the same result as clear-cutting.

Clear-cut forests may take centuries to recover. Not only is the soil deprived of plant nutrients, but the organisms that perform decomposition are unlikely to survive, as they are adapted to life in a moist, shady environment. When the forest is cut, sunlight floods the earth, heating and drying it and even altering the flow of water and minerals within the soil. Such changes make virtually impossible the survival of rainforest species that reduce dead matter to useful nutrients.

Plant Life of the Tropical Rainforest

Plants feed upon stars, or at any rate upon one particular star. By drawing in the sun's energy and converting it to earthly living tissue, plants bring the world of the stars into the world of our lives. It is a process that occurs in the leaves, where the green pigment chlorophyll captures sunlight and converts it to plant food.

For rainforest trees, life is an endless struggle for sunlight. The trees, at an immeasurably slow pace, race one another to the top of the forest, each trying to get its crown into the canopy, the great living lid of the forest. Every plant left behind must be very shade tolerant or it will be forever stunted or, just as likely, die. The tall trees dominate the forest; their height and their wide-reaching crowns give them a distinct competitive edge against all the other green plants struggling to live around them. The need for sunlight even shapes the trees themselves. As new branches grow from the treetops, they shade the branches below. The lack of sunlight starves and kills the lower branches and they fall away, leaving the trunk below the crown as stark as a Grecian column. As a result, the trees are shaped like giant broccoli, with all the greenery mushrooming at the top of the slender trunk.

From canopy to ground is a long drop in the richest forests. Rainforest trees average about 135 feet tall. The very tallest trees may soar fully 200 feet. These trees are called *emergents*. Their crowns pierce the canopy and tower over it, looking down upon the other trees like solitary sentinels, absorbing sunlight without competition.

The shallow roots of rainforest trees give them a tenuous grip upon the earth, so almost every storm that blows through the forest is punctuated by the crash of trees whose weary roots have given way. Many of the taller tree species have developed a peculiar adaptation that may serve to brace them up. These braces are finlike buttresses that rise up the trunks for as much as 20 feet. They look so much like buttresses that anyone seeing them is forced to conclude that they stabilize the tall trees and keep them from falling victim to the shallowness of their roots. But whether that is truly their purpose is still subject to debate in botanical circles. Some botanists believe that the long fins increase the amount of wood within the tree, allowing it to carry more sap. This increase in moist sap would be advantageous to tropical plants, which tend to lose a lot of water through transpiration, the process by which water vapor escapes through leaf pores. But if the fins were meant only to give the trees more room for sap, why are they restricted to the bases of the trees? Perhaps they serve a double function—both helping water balance and augmenting durability.

Most seeds produced by trees of the climax forest—the complex of species characteristic of a fully mature rainforest—are too heavy for wind dispersal. The pods of a Brazil-nut tree may weigh 5 pounds, while those of a monkey pot tree may weigh twice that; even the much smaller seeds of an avocado are roughly the size of golf balls. These trees rely on animals to disperse their seeds. Since rainforest animals may be reluctant to enter a big clearing, the climax species cannot move easily into the center of clearings, but must drift in languidly from the edge. The trees are therefore very slow to reclaim lost land. They may take centuries to recover what seems a small loss of territory. The large seeds have another disadvantage. They require a great deal of moisture if they are to remain viable, and the moisture makes them susceptible to fungal infection. Consequently, in many species the seeds lose their viability within a month.

But the heavy seeds are also vitally advantageous among trees competing for sunlight because their bulk is due to large quantities of stored energy, energy that the seedlings use for growth. Because they do not need sunlight to sustain them, the seedlings of some species can grow as much as 36 inches before sprouting leaves. The advanced height gives them a distinct advantage over low-growing shrubs and seedling trees that must produce leaves quickly.

It is difficult to determine how long rainforest trees live after passing

A view of the forest interior of Costa Rica's Braulio Carrillo National Park, 5,800 feet above sea level, shows dense growth. Rainforests, teeming with life, are home to about half of the world's species.

into maturity, because their growth rings do not reflect annual seasonal changes and yearly growth as rings in temperate-zone trees do. But it is suspected that if a tree lives into senescence, it may fill a place in the canopy for several centuries. In the end the tree will fall, opening a new swath in the forest, its trunk and crown dissolving back into the soil, becoming food for new trees.

Though trees are the dominant feature of the rainforest, many other plants live among them. Perhaps foremost of these are the lianas, woody vines similar to the Virginia creeper of North American forests and the woodbine of Europe. But rainforest lianas are much larger, with stems often growing to as much as 6 inches in diameter and well over 200 feet long. For example, the longest known rattan—a type of climbing palm—reached nearly 540 feet.

Lianas grow up the trunks of the trees and loop through their branches, an adaptation that allows them to reach sunlight. A single tree may bear three or four different lianas. The vines reach into the canopy, linking tree crowns to one another and providing support during light storms. In a heavy wind, however, the lianas may prove deadly, allowing a falling giant to drag down other trees to which the lianas bind it.

Some evidence suggests that lianas are important to the cycling of nutrients. Though the vines account for only 5 percent of the wood produced in a rainforest, they account for 36 percent of the leaves that litter the ground, their nutrients feeding the soil. They are also economically important. Rattan is heavily harvested in Southeast Asia, where nearly 500 species have been recorded. Rattan is sold worldwide for use in mats, window coverings, furniture, and many other objects.

The rainforests are also home to the largest herbs in the world. Herbs are plants that lack woody parts, though in many species the leaves form a sort of pseudo-stem. Rainforest herbs include the banana. Some rainforest herbs produce huge leaves. *Monstera*, a Mexican herb, produces leaves that are about a yard across and grow on stems a foot or more in diameter. Taro, an herb that grows in Malaysia and Indonesia and produces an edible tuber widely consumed there, sprouts leaves that are nearly six feet long.

Another important plant type is the epiphyte. These begin life as wind-borne seeds or as berries that are carried through the forest by birds. They grow when they reach some object that gives them a foothold, such as a tree branch. They never touch the earth. One familiar example is the commercial houseplant that dealers call *air plants*. The branches of some trees may be crowded with up to 65 different epiphyte species.

Perched above the ground without earthly contact, epiphytes have evolved some fascinating mechanisms for getting food. In their roots, some epiphytes harbor a fungus that is able to tap into the living tissue of the host tree and draw out nutrients that the fungus makes available to the epiphyte. Heavy infestations of parasitic epiphytes will kill a tree.

In other epiphytes, the leaves are shaped like funnels and catch debris falling into them from higher up the tree. The debris remains in the leaf, absorbing moisture and turning into humus, a sort of organic

paste from which the epiphytes can absorb nutrients. Some epiphytes bear hollow leaves that provide homes to ants. The ants feed on insects, bringing prey back to their vegetative caverns for consumption. Leftovers accumulate in the ant's lair, and the epiphyte's roots absorb this decaying matter.

Many epiphytes that live in tree crowns, where they are subject to rapid changes in humidity, have developed structures for collecting water. Leaves may be funnel or pitcher shaped, suitable for collecting and holding rain water. Large epiphytes—some bromeliads are more than six feet in diameter—hold as much as a gallon of water and function as miniature ponds in the forest crown. Birds, monkeys, and other animals get water from them, and frogs lay eggs in them. A single epiphyte may provide home and moisture to several species of insect and spider. When the live-in animals die, they become food for the epiphyte.

Some entirely parasitic plants are unable to produce their own food because they lack chlorophyll. They are particularly interesting because among them are the species that produce the largest flowers in the world. These 10 to 15 species belong to a group called *Rafflesia* and are found primarily in west Malesia (Malesia is a generic term for the area encompassing Malaya and Indonesia), where they grow on lianas. Their vegetative parts, which look like mere threads, draw nutrients from the host's tissue. Their flowers are gigantic, nearly a yard in diameter. The purplish flower of one Sumatran species smells like decaying meat, presumably to attract flies and beetles that help with pollination.

Tree branches provide growing sites for another type of plant, the

An epiphyte takes root on a tree trunk in Costa Rica's Monteverde Cloud Forest Biological Reserve. Bromeliads and other large epiphytes are specially adapted to changes in humidity and can hold a gallon of water within their leaves.

stranglers. The fruits of stranglers are eaten by birds and bats, which often drop uneaten seeds on the branches of trees. The seeds sprout, sending out a root that may grow for as much as 130 feet before reaching the ground. On the ground, the root rapidly branches, forming a complete net around the trunk of the host tree. The host may be strangled, dying away in the cage of roots. The tree eventually may rot away, leaving the strangler roots looking like an empty basket supporting the vegetative parts of the strangler and providing homes for many animal species.

Not only the branches, but also the leaves of rainforest trees and shrubs provide homes for other plants. Tiny lichens, liverworts, and some mosses grow on the leaves of larger plants. If a leaf becomes too heavily encrusted, it can no longer receive the sunlight it needs to survive. When it subsequently dies and falls from the tree, it carries with it the lichens and other life growing on it, and all the nutrients of this miniature ecosystem pass into the soil.

When the Forest Is Destroyed

Rainforests are destroyed by any of a number of forces. Fire, flooding, landslides, and clear-cutting all cause some level of destruction.

The process that turns cleared sites into primary forest may take hundreds of years. Forests cleared by the ancient Aztecs and Mayans before European discovery of the New World still have not reached the climax condition. Large clear-cuts in Cambodia have failed to revert to climax forest after 600 years of solitude.

The speed of succession in tropical rainforests depends on a number of factors. One is the size of the destroyed area. As mentioned earlier, the big, heavy seeds produced by many rainforest trees do not easily disperse into cut areas because of their limited viability and their dependence on animals for distribution. Moreover, forest trees do not reproduce until they are several decades old, another factor slowing forest recovery. Even a tree that produces relatively lightweight seeds would probably take at least a century to advance roughly half a mile, in part because all the interdependent plants of the rainforest must advance together, a process that moves at a speed of roughly 100 yards per century. But this can happen only if the cleared area is near a large expanse of untouched forest that has the full complement of species that characterizes a climax forest.

Even this scenario is overly optimistic because it does not account for the various obstacles to forest recovery. It fails to consider that the loss of the forest changes the conditions the forests need for growth. For example, about 75 percent of the rainfall in the Amazon Basin originates from moisture released by forest plants. If large parts of the forest are destroyed, rainfall may well decline. This may make it difficult, if not impossible, for surviving rainforests to reclaim cutover areas. It is more likely that the loss of moisture would cause declines in surviving uncut forests. The destruction could thus grow like a cancer, powered by its own pernicious force.

Today, rainforest destruction is alarmingly widespread and proceeding much too fast to allow the plants and animals to adapt. Within a few years, there may not be reserves large enough to ensure the survival of the rainforest: Protected forests in most nations constitute much less than 5 percent of the land area. Moreover, the lands that are protected tend to be fragments left by loggers or mountainous terrain that the loggers could not reach. The richer lower-level forests are the first cut and the last protected.

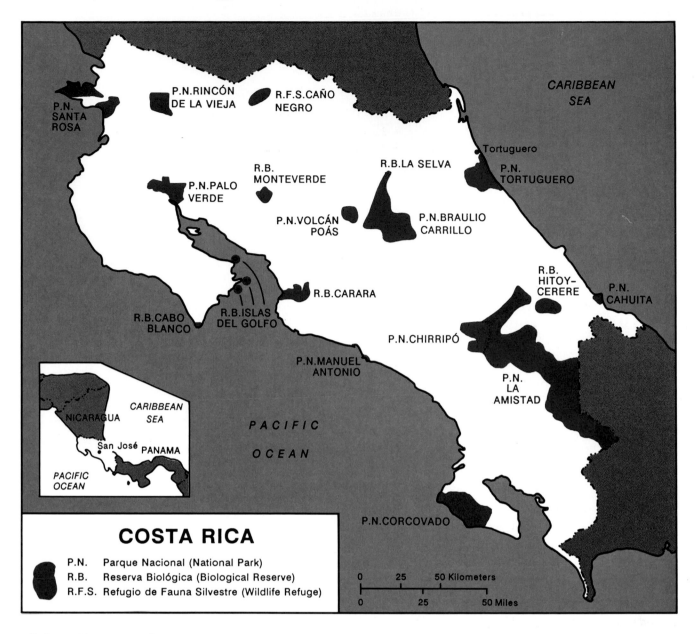

COSTA RICA

P.N.	Parque Nacional (National Park)
R.B.	Reserva Biológica (Biological Reserve)
R.F.S.	Refugio de Fauna Silvestre (Wildlife Refuge)

0   25   50 Kilometers

0   25   50 Miles

About 12 percent of Costa Rica, a country the size of West Virginia, is national parkland; another 15 percent lies in reserves and other parks.

Costa Rica's Experiments in Ecosystem Biology

Costa Rica lies in the southern end of Central America, just north of Panama. The first European to arrive there was Christopher Columbus, who landed near today's Port of Limon in 1502 and thought he was close to Siam. Delighted with the many gold gifts the native people gave him, he dubbed the territory *Costa Rica*, the rich coast. The first European settlement was Cartago, founded in 1560 by Spanish settlers. Cartago was built in the central highlands, a cool region surrounded by mountains and volcanos. Subsequent settlement of Costa Rica by Spanish settlers continued to focus on the central highlands, where the colonists were isolated from the outside world by the surrounding mountains. They lived by their farming and, losing touch with the homeland, became increasingly independent and democratic. When wars for independence from Spain set Central America aflame, Costa Rica continued to drift along in peaceful isolation. When independence was won in 1821, the Ticos did not hear about it until the news was more than a month old. With freedom assured,

Costa Rica joined the Central American Federation—a union of nations—but in 1838 declared itself an independent republic.

The old traditions of farming and democracy are still strong in Costa Rica. The *campesino*—farmer—remains a culture hero among the Ticos. Coffee, introduced to the rich volcanic soils of the central highlands in the early 1800s, and bananas, introduced at the turn of the century, are the foundation of Costa Rica's economy. The roots of democracy are deep. Costa Rica established free, compulsory, tax-supported schools in 1869, long before the United States followed suit. Nearly a quarter of the national budget goes to schools. Students wear uniforms to minimize the distinctions among children of different economic backgrounds. Costa Rica has one of the highest literacy rates in the Western Hemisphere. The social security system aids all citizens, and the infant mortality rate is among the lowest in Latin America. The death penalty was abolished more than a century ago, and there has been no army since 1948.

The colonial era had little impact on Costa Rica's rainforests. By 1900, only 75,000 people lived in Costa Rica, a nation roughly the size of West Virginia. The population center remained in the central highlands. But this has changed drastically in recent years. In 1946 the Pan American Highway opened south of the central valley and brought in thousands of settlers. By 1956 the central valley was home to a million people. Within 20 years, this figure doubled. Today, Costa Rica's total population is 3 million. As the population increases, so does pressure on the land. Only a third of Costa Rica's original forests remain, and nearly 40,000 acres of forest are still being cut yearly to make way for crops and cattle.

This loss of forest threatens the survival of many species. Costa Rica covers only about .0003 percent of the Earth's surface, but houses 5 percent of all known plant and animal species. This includes 850 bird species, 35,000 insects (probably a gross underestimate), 9,000 vascular plants, 208 mammals, 220 reptiles, 160 amphibians, and 130 freshwater fishes. Moreover, Costa Rica acts as a land bridge between North and South America. It is home to species from the north and the south as well as to species that exist only in Costa Rica. It is a species-rich meeting ground.

Costa Rica is struggling to protect its biological gold mine. It is a tremendous cultural challenge because of the nation's agrarian tradi-

Christopher Columbus lands on a Caribbean island in a 1921 painting by George Peter (1860–1945). The famous explorer was the first European to visit Costa Rica, which he named the "rich coast" because of the many gifts of gold offered by native people.

A scarlet macaw in Costa Rica alights on its perch. Though rainforests are full of life, many species are secretive or so well camouflaged that spotting them is a challenge. Birds of the dense forest are more rarely seen than heard.

tions. To own land and to be able to pass it on to descendants are vital aspirations among Ticos. Clearing land is part of their heritage. But despite this tradition, the Ticos are leaders among Latin Americans in their effort to protect the rainforest. About 12 percent of Costa Rica has been set aside in national parks. Another 15 percent lies in reserves and other protected areas. However, nearly all forest reserves and about 20 percent of the region encompassed by national parks are in private ownership because of inholdings, making long-term protection impossible to ensure.

As is true of other tropical rainforest nations, the long agrarian tradition has made it essential that Costa Rica try to find ways to save its rainforests by using them prudently. Unless the forests produce some benefit for the Ticos who live near them, the forests will never be safe from the inroads of campesinos. Walking the fine line between exploiting the forests and protecting them is the challenge that the Costa Rican government and Tico biologists have set for themselves.

Taking Account

The first step in attempting to protect the rainforest is to determine what species it shelters. Until biologists know precisely what needs to be protected, all efforts will be essentially without clear goals and objectives. The second step is to determine ways in which native species can be used wisely for economic gain.

The task of assessing what lies in the rainforests is headquartered in two battered warehouses at the end of a dirt road outside San José.

This is INBio, the National Biodiversity Institute of Costa Rica. Inside the front door of one of the routine-looking warehouses lies a routine-looking reception office—neutral walls, bare floors. The only thing that suggests this is not your average office is the variety of mounted birds of prey that stare with glass eyes from perches on the walls.

Beyond the reception office INBio shows the stuff of which it is really made. At a table in the middle of a large room lined with cabinets, a slight, dark-haired young man intently studies an array of insects that have been impaled with pins and mounted in a shallow box. The young man is Reynaldo Aguilar. He has been collecting insects, other animals, and plants for INBio for the past year. He is a parataxonomist, a new breed of biologist created and trained by INBio to help with the critical task of cataloging the nation's rainforest species.

A one-time factory worker and auto mechanic with a passion for nature, Aguilar worked as a park guard at Braulio Carrillo National Park for five years, including three years as an armed guard, before joining INBio. He went through special training at INBio to qualify as a parataxonomist. Now much of his time is spent collecting specimens at the park he guarded, a task that he enjoys. "I like Braulio Carillo, and every day I like it more because it is the rainiest of the forests and it's been studied less than the drier forests," he says.

Aguilar's initial interest was in birds, because they are the most easily observed of forest species. But his allegiance soon switched to mammals, particularly bats and rodents. After he was trained at INBio he became interested in insects. He learned that bats eat a certain insect that is a parasite of birds. This taught him about the interrelationships that exist between species, even those as distantly related as mammals, birds, and insects. The trees, too, fascinate him. "For me, trees are a work of art," he says. "They are full of gold. We need to learn to use them. The more we learn about them and their biology then the better we'll be able to use them."

The science that Aguilar has learned at INBio has also helped him with his work as a park guard. In the past, farmers living near the park would come to him with questions he could not answer. Some who had lived in Braulio Carrillo wanted to know why they had been forced to leave their land when it was declared a national park. It was difficult, he says, to tell local residents that they had to stay out of the park and leave it alone. Now, he says, he knows what he is guarding and why it must be protected. Plants in the forest can provide medicines, he says, and when he explains this to the local farmers, they too begin to understand why they must stop clearing national parks. Aguilar says that explaining the value of the forest in terms of medicinal plants is a happier solution than trying to use weapons to keep people out of the parks.

Clearly, protecting the forests requires the support of local residents. Residents often do not understand the concept of a park, but by showing them how their activities have wiped out certain fish species or peccaries—a species of wild pig—Aguilar is able to teach these people about the consequences of their actions. When they realize they are losing species that they need for food, they begin to understand that park protection is important to them as well as to the rainforest.

Because of his training, Aguilar is able to speak about plants and

A white-faced capuchin monkey hangs loose in Costa Rica. The forests and wildlife of this small Central American nation are besieged by poachers, farmers, ranchers, and lumbermen. Biologists hope to document as many species as possible before they disappear.

animals with scientific authority. But because of his familiarity with local people, he is also able to communicate with them in terms they understand. This allows him to glean a great deal of information from them that more technically trained, academic biologists might not be able to solicit. Academics, he says, know the scientific names of various species, but the locals tend to have more knowledge and understanding of the plants and animals in their domain. However, they tend to be reluctant to share this information with scientists because academics intimidate them.

Reynaldo Aguilar is one of some 30 parataxonomists trained at IN-Bio. The program began in late 1989, the result of a meeting earlier in the year at which researchers from various institutes discussed ways to better integrate the work they were doing. Many biologists were collecting data in isolation from one another, and the specimens they collected were treated like personal possessions. Information was not being shared. At the February meeting it was determined that the Ticos themselves, rather than North American and European biologists, should be collecting specimens. In June 1989, President Oscar Arias created a special committee to organize INBio, and by October the group was under way.

The director of INBio is Rodrigo Gamez Lobo. His program has been an amazing success. Within the first year, he said, it became necessary to slow down the field parataxonomists. They were collecting so many specimens—100,000 per month—that those doing the cataloging at headquarters could not keep up. The rate at which specimens accumulated was unprecedented in the annals of science. Traditionally, collections were gathered by big museums in Europe and North America. University-trained experts would make expeditions to the rainforest, spend some time collecting specimens, and then return home to study the collection for several years. This was a very slow approach, and the collection tended to be commandeered by the collectors. The nations whose species were being studied gained little from the research.

The use of parataxonomists at INBio has accelerated the process. The staff collected more material in eight months than the traditional system had during the past 110 years, Gamez says. In many ways, he says, local residents are better adapted to the task than are outside experts because they are more accustomed to the heavy rains, insects, snakes, and other difficulties inherent in forest work, and they are willing to collect both day and night.

The parataxonomists' inventory of rainforest organisms will provide essential information for determining how rainforest protection must be managed to avoid the loss of thousands of species. But they are up against two relentless opponents. One is the sheer number of species being inventoried—perhaps half a million, perhaps much more. The other is the speed with which the forests are being lost. If conditions do not change, Costa Rica will lose most of its rainforests by the end of the decade. "We are fighting a big battle to protect and save biodiversity," says Gamez.

The parataxonomists come from various walks of life, including housewives, ranchers, park guards, farmers, policemen, and field workers. All live in the areas in which they make their collections. Because of Costa Rica's advanced school system, INBio has at its disposal a citizenry that has the basic education needed for more specialized training. At INBio, they learn the fundamentals of ecology, biology, and specimen collecting and mounting. They also get advanced zoological training in particular groups of animals. The course takes five to six months to complete.

Work is supervised by experts from such institutions as the Smithsonian and the British Museum. The experts monitor the collections and offer advice about which species need to be more thoroughly collected. This, too, is a new approach. The experts come to Costa Rica not on expeditions, but to interact with local people. The experts benefit because they get a box of beautifully prepared specimens and can publish any data that they glean from their work.

The techniques used to collect specimens vary with changing environmental conditions, such as which flowers are blooming and whether the moon is full. To catch insects, the parataxonomists may use traps baited with meat, manure, or flower scents. For night-flying insects, they use light as a bait. In parts of South America, entomologists spray a fogger into the trees and collect the insects that fall. This does not work in Costa Rica because winds frequently blow.

The parataxonomists take notes on their specimens, recording such details as where and when the plant or animal was collected. They also record certain peculiarities, says Gamez, such as, "I collected this plant and it had no insect bites.'' This might indicate that the plant produces a natural pesticide that could have economic value.

The specimens arrive at INBio headquarters fully mounted. At headquarters they are cataloged and filed in drawers. Each specimen bears a bar code so that information can be instantly called up on a computer. Gamez's goal is to place 200 parataxonomists in the field and eventually to expand their work to the collecting of marine specimens and even microorganisms.

Gamez sees the inventory of rainforest species as crucial to saving the rainforest itself. At present, most Ticos see the forest as valuable only for producing timber or providing land for agriculture. As long as the intact forest lacks meaning or value to local people, they will continue to destroy it. The inventory will determine the numbers and kinds of nearly all species present and what their potential economic value may be. This will help reveal many new ways in which the forest can be used for economic gain.

Another project aimed at both using and saving the forest is also under way in Costa Rica. It is being run by the Portico Company, which specializes in making fine doors from tropical woods.

The Portico Plan

Portico, based in San José, Costa Rica, turns out some 80,000 luxury doors yearly for export to the United States, Canada, and Europe. In 1985, the two-year-old company was on the verge of shutting down because, on lands then available to the company, deforestation by Portico and others had eliminated the trees needed to produce the doors. Now, business is booming, and the company employs 850 people. The turnaround was founded on a complete change in the company's policy concerning the cutting of trees.

The change began with the hiring of several forestry experts. They examined the company's logging policy, which called for clear-cutting vast areas of forest as part of the process of extracting the few fine-wood trees the company needed. Instead, the foresters said, the company should cut only mature trees of the species needed. Over a period of years, the forest would replace the trees, providing an endless supply of wood.

To adopt this policy, Mario Barrenechea, the president of Portico, needed to buy up vast acreages of forest before it was destroyed for agricultural development or other projects. Financing was a problem, however, until Ron Wynn of the Northwest Bank of Minnesota stepped in. Wynn had a problem of his own in Costa Rica. The nation was not making payments on a $6 million loan his bank had extended to it. Wynn formed a partnership with Barrenechea and struck a deal with the Costa Rican government: The debt would be scrubbed from the books in exchange for 15,000 acres of prime tropical rainforest, which would be managed for a sustained yield of wood for Portico's doors.

From his well-appointed office, Mario Barrenechea explains his company's logging policy. "Our intention is that, through the management process, we'll promote survival of most of the young trees," he says. "Everything indicates that after the first three or four cutting cycles, the overall volume of commercial standing trees will be greater than it is at present."

The logging that is supposed to bring about this desirable result involves selectively cutting only mature trees that are more than 70 inches in diameter at breast height. Such trees are no less than 75 years old, and may be as old as two centuries. The plan is to cut only two or three of these trees per acre every 15 years. Young trees are left to grow. Since average growth is roughly one-half to three-quarters of an inch in diameter yearly, the population of trees between 10 and 60 years old is more than enough to sustain the harvest at its present volume through three or four 15-year cycles. After that, sustaining the trees will depend on regeneration and new seedlings.

Portico is currently cutting nearly 2,000 acres yearly, taking out some 2,800 selected trees. The cut trees can be replaced within 10 to 20 years. About 85 percent of the trees used by Portico come from its own land. This was not always the case. When Portico opened, it bought wood from local markets. The quality of the wood was often bad and the supply irregular. In a short time, the supply diminished because of overcutting, threatening the company's survival. In 1987, Portico initiated selective logging on its own property, solving the quality problems and, it is hoped, establishing a means for a sustained supply. The vagaries of weather, muddy roads, and truck mechanics still force the company to buy about 15 percent of its wood from sawmills, but Barrenechea hopes to eliminate that element.

Portico's acreage lies in northeast Costa Rica, one of the rainiest regions in the nation. It is flat, low land. To get to it, you have to travel by four-wheel-drive vehicle across dirt roads flanked by banana plantations, over rocky roads that jar every bone, and, finally, depending on the season, across an hour or two of mud roads that will bog down even a four-wheel-drive with impunity. Then, with howler monkeys roaring from one of the few strips of trees left in an area largely cleared for farming, you slog on foot through the mud to a gate where Portico's forested land begins. At that point, travel really gets rough.

You soon find yourself out of the sunlight as the forest closes around you. As you sink into the shadows, you drag yourself through the knee-deep mud where a bulldozer has cut a road through the trees. The mud sucks your boots off your feet, sweat soaks your shirt and drips—no—*streams* from your brow. You see fallen trees with trunks so thick that even as they lie on the ground you can barely see over them. You continue down the trail, through standing water, and finally come to a leaf-strewn trail that roams along under the friendly shade of the trees.

Some of the trees are painted with bright yellow numbers. This is a code that indicates which trees will be cut. Each number is recorded on a map. When the time comes to cut, the loggers try to determine in which direction the tree should fall to minimize damage to other trees. They try to keep gaps created by falling trees to less than 1,500 square yards. The average gap is about 500 square yards. Sometimes, if felling a big tree would create too big a gap, the loggers simply spare the tree. The tractors that haul out the logs are carefully restricted to

Rodolfo Peralta, forester for Portico, checks selectively cut trees in northeast Costa Rica. The trees take 20 to 100 years to reach maturity.

narrow trails, called skid trails, to minimize damage caused by roads. After the cutting, with the exception of the small gaps where broken tree stumps rise above the rubble of lesser plants, the forest looks untouched. Most wildlife will hardly be affected by the loss of the trees. This is a vast improvement over the fate that came to most of the forest surrounding Portico's land—conversion to open pastures where cattle graze over sparsely vegetated red earth. "The reason we're doing this is, in the end, a business reason," says Barrenechea. "It's not a romantic reason, though sometimes it sounds like we're a little crazy, but, in the end, it's a business reason. The ultimate reason we're doing it is because we want this forest to survive in perpetuity, because that assures the long-term viability of our business."

But will it work? No one knows for sure. The techniques being used by Portico were developed for forests in the northeastern United States. It is not clear whether they will work in the tropics. In any event, long-term commercial management of tropical hardwoods on the scale that Portico is attempting it has never been tried in Costa Rica. Also, the roads that Portico cut into the forest have, in the past, served as avenues for trespassing farmers who subsequently destroyed any forest that remained. Moreover, Portico is planning to remove rare species that exist nowhere else. The outcome is uncertain.

But studies are being done to determine how the new logging policy works. Biologists are conducting before-and-after inventories of wildlife on Portico land so they can assess damage to the ecosystem. The company is under close scrutiny by the Costa Rican government and international conservation groups. In hopes of proving its critics wrong, Portico is allowing an independent study to take place on its property. Participating in the study is Mike Graham, a botanist with the Missouri Botanical Garden. He says, "This is an experiment, and it's an ecological experiment, and it's an economic experiment. It's an economic experiment for Portico, and they're taking kind of a risk, more or less putting their necks on the line. But it's an experiment that, whether it succeeds or fails, will be very valuable to the whole tropics all over the world."

Jim Barborak of the Caribbean Conservation Corporation, who serves as advisor on a plan to create a rainforest reserve that will span the border of northeast Costa Rica and southeast Nicaragua, says the jury will be out on Portico for 10 years or more, until the second cutting cycle. His main concern is how Portico, whose land overlaps the proposed reserve, will affect the ecosystem. The project calls for the reserve to run from Tortuguero National Park north across the Rio San Juan into Nicaragua, joining Tortuguero with Nicaragua's newly created Rio Indio-Maiz Biological Reserve. The plan for the consolidated park came out of a meeting of Central American conservationists in 1974. Civil war in Nicaragua stalled the plan for 15 years. "It is only in the past two years, as peace broke out along the river, that the project has finally gathered momentum," Barborak says. The two-nation park would protect nearly a quarter million acres in Costa Rica alone, plus three times as much in Nicaragua. Because of warfare in Nicaragua, rainforest destruction was slowed. Every time a sawmill went up in the southeast, says Barborak, it was blown up and the technicians kidnapped. The violence helped protect one of the finest stands of rainforest north of Colombia. The joint operation should benefit both nations. Nicaragua, Barborak says, has about the same population as Costa Rica but is nearly three times as big, with more intact rainforest. But the Nicaraguans lack the enforcement personnel needed to protect the forest. The Ticos have less forest remaining, but are developing a system for protecting it. The combination should be productive.

Portico owns several thousand acres in the center of an area that Barborak and other conservationists hope to use as a corridor linking Tortuguero National Park to lands to the north, on the Rio San Juan. Barborak fears that logging in the center of the corridor could affect its use by wildlife. Nevertheless, he sees Portico as a big improvement over more traditional uses of rainforest lands—cattle grazing and banana plantations. Having Portico land adjacent to the park, he says, means roughly 40,000 more acres of forest for most species, since Portico seems likely to leave its forest largely intact. "Almost everybody would love to see Portico succeed," he says, "as long as its success does not threaten the long-term viability of the neighboring park."

Portico is a unique project that may provide useful information for others seeking ways to protect the forest by using it. Clearly, Portico is attempting to protect the ecosystem only because of self-interest: Without the forest, Portico would vanish. A similar attempt to give people an economic incentive for protecting rainforest is being formulated on the west coast of Costa Rica on a ranch of sorts that is the inspiration of a single determined woman.

Iguana Mama: Lizard Meat, Anyone?

She calls herself the Iguana Mama, and there is some justification for the title. German-born herpetologist Dagmar Werner's iguana-breeding project has made her mama to something in excess of 30,000 of the big lizards, which grow up to 7 feet long.

Werner's iguana ranch is tucked away on a farm outside Orotina, a village on Costa Rica's Pacific coast. The lizards are kept in various types of enclosures, ranging from kennel-sized wire cages to room-sized corrals of concrete and wire mesh. The iguana stock is capable of producing as many as 13,000 eggs during a single breeding period. The eggs are hatched in incubators and the young reared in cages until they are seven months old. Then they are distributed to farmers, who release them on their land. The farmers may either provide the lizards with food or let them fend for themselves. The lizards live in trees and feed on leaves, so trees are essential to the survival of the lizards. Getting the farmers to raise the iguanas is thus linked to tree protection: It gives farmers a reason for saving trees.

But what gives farmers a reason for raising lizards? The answer is food. The green iguana has graced local dinner tables in the area we call Central America for some 7,000 years. Many rural poor depend on the lizards for meat, and the eggs are reputedly used as an aphrodisiac. About 30 percent of the people in Panama and Nicaragua eat iguana meat. Werner believes that about 70 percent of Panamanians would do so if the meat were more available. Her goal is to make it more available.

Iguana meat is white and tastes much like chicken, a fact that was no doubt instrumental in bestowing upon the iguana its Central American nickname, chicken of the trees. The green iguana could provide protein to people living throughout its range—from southern Mexico to Brazil and Paraguay. Males may weigh in excess of 12 pounds, through 8 pounds is more typical. Females are about half that size. The animals grow slowly. A chicken can grow to 6 pounds in about four months. To reach the same weight, an iguana takes four years. But the iguana needs 30 percent less food than does a chicken, so economically it is more practical in the long run. As a native of the tropics, it is better adapted for life there and can take care of itself with minimal human intervention. The meat it provides could be indispensable to tropical farmers in remote areas. These farmers usually cannot afford to buy beef, and lack the refrigeration necessary for storing large quantities of meat, even if they raised their own cattle. In any case, iguanas—as well as armadillos and the large rodents called agouti and paca—are a better source of protein. But iguanas require trees, which are usually the first things to go when farmers move into a forest.

Both the hunting of green iguanas for meat and the felling of forests have jeopardized the lizard's survival. Twenty years ago, whole truckloads of iguanas were taken to markets in various Central American nations. But now the animal is listed as endangered in Panama and is extinct or vanishing throughout most of the rest of its range, retreating as the rainforest retreats. In Mexico's Pacific Coast mangrove forests,

At her Costa Rica lizard ranch, Dagmar "Mama Iguana" Werner holds 8-year-old Ignacio, whose brand is visible on his side. Werner has played mama to more than 30,000 green iguanas since 1985.

iguana populations are down 95 percent. Only remnant populations are left in Guatemala and El Salvador. The price for an iguana has risen from 80 cents in 1976 to about $20 on today's black market, a sign of how scarce the animal has become.

The idea of raising iguanas in captivity dates back to the 1960s, when the Smithsonian's A. Stanley Rand began studying green iguana behavior. He found that female green iguanas migrate en masse to nesting sites along the sandy banks of rivers and lakes and along the seashore. They lay their eggs in deep, elaborate burrows. The hatchlings tend to remain together in reptilian crowds. In the 1980s, destruction of iguana habitat led Rand to conclude that survival for the species might depend on captive breeding. The group living arrangements of the green iguana made it a good candidate for captive breeding, and animals reared in cages could be released to bolster wild populations.

The breeding project got under way when the Smithsonian hired Dagmar Werner to run it. Werner had just finished her doctorate after spending seven years in the Galapagos Islands studying land iguanas and lava lizards. She understood both the lizards and the critical role that campesinos play in any effort to protect wildlife.

Werner, who peppers her conversation with liberal doses of boisterous humor, started her work in Panama in 1983. She captured several dozen pregnant iguanas and released them into a 100-square-yard pen that enclosed a natural iguana nesting site outside Panama City. The lizards dug their burrows and laid 30 or 40 eggs each, but Werner discovered that iguana burrows are so convoluted that digging them up to collect the eggs was almost impossible. She and an assistant had to dig up the entire pen three times to collect 700 eggs. She hatched the eggs in an apartment in Panama City and housed the hatchlings in cages made from bamboo, natural vegetation, and metal sheeting of the sort used in roofing—all materials readily available to the campesinos to whom she hoped one day to turn over the lizards. About 95 percent of the hatchlings survived, a marked contrast to life in the wild, where predators and other threats wipe out all but 5 percent of any given hatch.

To make the collection of wild eggs easier, Werner buried artificial burrows, made of concrete or clay tubing and cinder blocks, at natural nesting sites. The lizards, apparently recognizing a good thing when it was handed to them, used the prefab burrows rather than digging their own. Werner was able to collect the eggs with ease. By 1986, she was pulling 2,000 eggs per season out of the burrows.

Meanwhile, life was moving apace at the breeding center, which had been established at Summit Hill, a recreational park outside Panama City. In 1985, the first generation of iguanas reached breeding age. Two years later, Werner had 1,400 hatchlings from captive-laid eggs and stopped collecting wild eggs. By 1988, she had more than 8,000 green iguanas, mostly from captive-laid eggs.

Central to Werner's program is distribution of the iguanas to farmers. She began doing this in Panama in 1986, giving out some 1,200 iguanas in two rural communities. She initiated the releases with great fanfare, including a blessing from the local priest, in an effort to discourage poaching. During the next three years she distributed nearly 7,000 more iguanas in six communities.

The iguana project has been received with enthusiasm among campesinos. The project offers the possibility that a traditional source of meat will be resurrected. It also promises to generate some income: Each iguana should generate a profit even if sold for only $3, about a seventh of the black-market price. Also, iguanas are a cost-efficient meat source. Werner believes that 2.5 acres of forest turned into pasture will yield about 33 pounds of beef yearly, but that the same land, left in forest or replanted with trees, could yield more than 300 pounds of iguana meat each year.

If campesinos can be shown that iguana production is more efficient and profitable than cattle grazing and crops, they will be likely to plant trees in cut-over areas and to spare forests in areas yet untouched. The

Baby green iguanas go bonkers for mango at Dagmar Werner's ranch. The males can grow up to 7 feet long and weigh 12 pounds. They will be distributed to farmers as a food source and an incentive to protect the forest.

green iguana, with its dependence on trees, may show Central Americans that humanity benefits directly from forest survival.

Panama's political woes forced Werner to uproot in 1988. She accepted a teaching job at Costa Rica's National University and moved her project to Orotina. Communities in Costa Rica, Honduras, and Guatemala are becoming involved in her work. Ticos, Werner says, do not traditionally eat iguana meat, but she believes that some Tico farmers will raise the animals for export.

The cooperation of the campesinos is essential to the success of the program, and part of the project involves teaching farmers not to hunt down every iguana that reaches breeding, and eating, size. Some must be left to provide stock for the future. The campesinos are beginning to understand the wisdom of this policy. In one Panama community, iguana poachers were arrested and required to work in the iguana project so that they would learn its value.

The cooperation of local residents is always a crucial factor in rainforest protection. One project in Costa Rica focuses strictly on the challenge of earning the cooperation of local farmers in rainforest protection.

Boscosa

The Osa Peninsula lies in the southeast corner of Costa Rica, near Panama. Some 100,000 acres of the Osa lie within the boundaries of Corcovado National Park, an unbroken expanse of prime wet tropical forest. Only a few footpaths cut through the rainforest, and those are used by only the hardiest travelers. Rainfall approaches 240 inches yearly on the highest peaks, with an average of 140 inches for the entire park. In the park you can count some 500 tree species, with an average of 100 species per hectare (about 2.5 acres). You may also tabulate some 367 bird species, 500 mammal species, 117 amphibians and reptiles, 40 freshwater fish, and more than 6,000 insect species.

The park, which is protected from development, is surrounded by 800-square-kilometer Golfo Dulce Forest Reserve, in which some types of development are allowed and which some 5,000 families claim as home. These settlers, often without legal permits, are in the process of clearing the reserve. The process follows a certain pattern, explains Alvaro Umana, vice-president of the Costa Rican Center for Environmental Study and former minister of natural resources. The settlers cut trees with axe or chain saw and sell the trees to loggers, who promise to build access roads into the forest so the settlers can transport goods into and out of their home regions. The roads make it easier for more settlers to move in. Meanwhile, any trees left by the loggers are burned with remaining forest. All told, about 80 percent of the wood never gets to market.

The settlers put the land into crops, but the soil soon washes away. As the soil erodes, so does the level of crop production. When crop returns reach an unprofitable level, the fields are turned into cattle pasture. Cattle compact the soil, further degrading the land. Eventually, the campesinos move on to new forests.

Until recently, clearing forest was considered an improvement to the land that earned the campesino title to it. If a settler, as a squatter, cleared land belonging to someone else, the settler could be evicted, but if a year had gone by the owner was expected to compensate the settler for the improvements made. This practice dates to the colonial era, when the Spanish crown assigned huge land grants to favored people. Often the recipients of the grants did not know precisely the boundaries of their lands, so, in general, if no one protested a squatter's activities, the land was considered ownerless. The effect of this situation was to reward campesinos for destroying forest and eventually led to the destruction of most of Costa Rica's rainforests. At the turn of the century, most of the nation was covered with forest. By the 1970s, forests were being burned and felled at an estimated rate of 150,000 acres yearly. Only 17 percent of the forest remained by 1983, and until recently roughly 75,000 acres were lost each year. Government figures indicate that logging has slowed to about half that amount yearly, but even at that rate most private forest lands and reserves will likely be exhausted within the next five years and pressure will build to open national parks to clearing, says Alvaro Ugalde, superior director of the Department of Wildlands and Wildlife in the Ministry of Natural Resources, Energy, and Mines and a founder of Costa Rica's 14-year-old national park system.

The cooperation of local residents is essential to protecting forest reserves and national parks. Forest managers, says Ugalde, must become more involved with the people who surround parks. "That's critical, because if at the frontiers people are poor and getting poorer, there is not much future for the park," he says. He believes that giving people title to the land they have claimed and teaching them how to use the land without cutting it are two factors that will help save rainforest.

These two factors are being put to work on the Osa Peninsula in the Boscosa Project (Boscosa is a combination of Osa and Bosque, spanish for woods). Boscosa is a pilot project that seeks to maintain forests on the Osa by teaching campesinos to reforest cut areas and to use agricultural lands intensively. Instrumental to the success of this project is Richard Donovan, a U.S. forestry expert hired by the World Wildlife Fund in 1987 to serve as the Boscosa Project's coordinator.

Donovan is a robust man who hikes the knee-deep mud of mountain trails with vigor and approaches his work with noticeable glee. He is at home in the Osa, where towns are often little more than a few houses and a general goods store and horses constitute perhaps the most common form of transportation. He hails from a logging background himself. "I have a family background in which my grandparents and my great-grandparents were involved in basically deforesting northern Minnesota. They did the same thing a lot of foot loggers are doing here—which is claiming it. They take the best trees and, to a large extent, they're not worried about what is left afterwards. And my grandfather and great-grandfather did a lot of that in northern Minnesota. We can't be sitting up there in the states, holier than thou about rainforests. We've done the same damn thing. We made the same mistakes."

Donovan spent three years in the Osa trying to teach campesinos not to make the same mistakes his own forebears did. He did this by teaching settlers ways to exploit the forest without wiping it out. "I really don't like to call it saving the rainforests," says Donovan. "I just like to call it that we're trying to mix the people in the forest in such a way that the forest stays and the people stay."

One way in which Boscosa accomplishes this is through a reforestation project. Local campesinos grow trees in nurseries, then plant them on cleared lands. Most of the trees—80 to 95 percent—are native species, a departure from most reforestation projects, which use eucalyptus and other fast-growing exotic species. The trees are to be harvested for lumber in later years. To take full advantage of the lum-

A homesteader lays claim to a hunk of turf in the Osa Peninsula of Costa Rica, near Corcovado National Park. Settlers plant crops on cleared lands until the soil is exhausted, then run cattle where crops no longer grow. More forest is then cut to compensate for the lost cropland.

ber, the Boscosa Project helps local residents to build sawmills and even furniture workshops for local manufacture of wood products.

The Boscosa Project also seeks ways to intensify agriculture, such as encouraging the use of perennial crops, those that do not need replanting each year, and developing better means for producing cocoa and beans on flatlands. The project also attempts to find ways to exploit native rainforest species. Project personnel worked with the 375 campesinos of Rancho Quemado to develop alternatives to land clearing. The project came up with 40 agricultural alternatives and nearly 30 other profitable approaches to forest management, including ecotourism, management of agouti and paca as forest meat species, production of ornamental plants, and even crocodile farming.

The theory is, says Donovan, that if you give people a stable agricultural buffer around the forest or make forest management productive, you keep pressure off the forest. If campesinos can be taught to maximize production from their land, then they will clear less land — or, as Hugo Alvarez, administrator of the Golfo Dulce Reserve, puts it, "The idea is to intensify production on farms so that the people are too busy to push forward on park boundaries." Also, if local residents can be shown ways to make a profit from native forest, they will want to protect the forest.

Donovan, who turned over administration of the project to Ticos in 1991, has won the support of many local residents. One is Jeremia Urena, a local preacher and campesino. Recalling his early days on the Osa, Urena says, "In six years we cut down approximately 203 hectares. That was because we had no chain saws. We couldn't buy chain saws. If we'd had chain saws, not even the park would have remained." Urena was the first local resident that Donovan approached. Converted to the idea of forest management, Urena bailed out of slash-and-burn agriculture and landed in the business of growing trees. He plants fast-growing balsa for lumber and peach palms that deliver fruit and heart of palm within three years. Replanting for long-term yields takes the pressure off the parks and provides a future for his family. "The forest was ruined because our thinking was such that we thought we had to discard, burn, clear," Urena says. "That's what we were taught by our fathers. Now you can see the landscape has changed. The whole area has changed. Now it hurts us to the very core of our soul—the gravity of the crime. But what's dead is dead. Now we have something to do. We've started now."

Part of the Boscosa plan calls for selectively cutting trees in the forest reserve in much the way that Portico takes out selected trees. It may seem ironic that the World Wildlife Fund's Richard Donovan occasionally directs the cutting of rainforest trees, but it is all part of a plan to protect forests. "We can harvest trees without upsetting the complete ecological system, and that can provide an economic livelihood for the people who live here, because right now every colone, every dollar that we pull out of the tree reduces pressure to completely clear the land," says Donovan. "And that's the direction they're going. There are other things that are easier. Cattle is easier. But some people like working in the woods, working with us on things that will last a long time, that are not just good this year, next year, but for a long period of time."

Changing local attitudes is part of the Boscosa Project's goal. The nurseries from which trees come for reforestation have won over many campesinos. Some 60 former slash-and-burn farmers make up the membership of one cooperative nursery. Said one farmer, "Before I started working with this company, I worked clearing the land. I felt very bad about it, but now I'm trying to repair the damage that I did. I feel that if someone doesn't concern themselves with doing this, our children will grow up never knowing what trees are."

At this stage, Boscosa is only a pilot project, a potential model for saving rainforests elsewhere. More people move into the Osa every day, and overpopulation fuels deforestation. A feasible economic plan that deals with this influx and that reconciles forest conservation and economic development is still on the drawing board. Meanwhile, Boscosa promotes forest management as a stopgap measure, a form of damage control. More money and more incentives are needed, not only to buy time, but also to invent the ultimate solutions.

"Tell me the answer to deforestation," says Donovan. "Anybody got it? I don't think anybody has it. There's no recipe here. And that's why we're trying lots of different things. . . . If we don't start showing them how to manage this forest for wood, my feeling is that the forest will disappear. Fully."

The Future

Nearly half of the 4 billion acres covered worldwide by tropical forests at the start of the twentieth century have been cleared. Every year loggers, farmers, ranchers, and developers in some 45 nations scattered along the equator cut down another 25 million to 50 million acres, and the rate of cutting increases year by year. By the second quarter of the next century, perhaps before, all but a few withering fragments of the world's tropical rainforests, along with the potential crops, medicines, fuels, and other resources the forests offer, will probably be gone.

The rate of cutting varies from region to region and nation to nation. Official figures on logging show that in southern and Southeast Asia, about 8,500 square miles of previously untouched rainforest are logged every year. In tropical Africa, the rate is close to 3,000 square miles, and in Latin America about twice that. Unrecorded, illegal logging could increase each of those figures by half, or perhaps even double or treble them. To these figures must also be added the acreage lost to agriculture and other development. Cautious figures from the World Bank's Economic Development Institute suggest that about 40,000 square miles of rainforests are lost to all causes every year, roughly a 1 percent loss annually. Biologist and rainforest expert Norman Myers and experts at the National Academy of Sciences have estimated the annual loss at about 80,000 square miles, or 2 percent each year. Even if you average these figures, the yearly loss is enormous. About 150 acres were brought down in the time it probably took you to read this page. Even at the low estimate, the rainforest will be exhausted within a few decades. Many of us will live to see the end of them, if current trends do not abate.

Costa Rica may hold some of the keys to saving rainforest not only in Central America but across the world. Yet even in Costa Rica, ominous signs cannot be ignored. Illegal logging continues apace. About 10 percent of national park lands are in private hands, and purchasing the land could cost more than a billion colones. Park protection has improved, says Alvaro Umana, but lands outside of parks are deteriorating. Many outside areas will be lost, and some park lands will vanish too. "You can't protect a park with guards and fences," he says. "The only protection is an educated, aware, and healthy people."

Alvaro Ugalde says only about half the national parks are properly managed, that poachers, gold miners, and campesinos are encroaching on park borders.

Conventional wisdom in Costa Rica holds that the battle to save the forests will be won or lost within the next decade, perhaps the next five years. The battle to save Costa Rica's rainforests, and indeed those of the world, is a fight that no one can ignore. Winning will depend upon the commitment of all nations, since the tropical nations cannot be expected to pay the bills themselves. But the outcome of the battle is important to all citizens of the globe. As INBio's Reynaldo Aguilar succinctly put it: "Humankind cannot live without the forest. If the forests are eliminated, humankind will be eliminated. That's what I believe. Therefore, we have to fight to protect our forests. We have to protect those animals who can't defend themselves. We are never going to see a group of monkeys going on strike saying, 'Hey! We want you to protect our forests.' Never. We have to do it for them."

TRIAL BY FIRE

On the afternoon of June 23, 1988, a thread of bluish smoke wafted from a pine forest a mile south of Shoshone Lake in Yellowstone National Park. George Henley, watching from a fire lookout on Mount Sheridan, spotted the smoke and, at 2:58 P.M., reported it to park officials. Smoke and small fires are common in summer in Yellowstone, so no one was alarmed. As part of a routine procedure, a park ranger hiked to the source of the smoke, arriving at eight that evening. He reported that the fire—dubbed the Shoshone Fire after the nearby lake—was small, covering only about 500 square feet in a stand of dead lodgepole pines, and surrounded by bare dirt. Yellowstone's Fire Committee concluded that the fire should be left alone to burn itself out, with two on-site personnel assigned to monitor the fire, weather, and other elements.

This hands-off approach to fire was part of a National Park Service policy that had been in effect at Yellowstone since 1972. National parks are among the most pristine of federal lands. Human activities are closely monitored, and activities such as oil and gas development, mining, and even hunting are rarely allowed in any parks. Unlike national forests and lands administered by the Bureau of Land Management—most of which are heavily exploited for timber, grazing, minerals, oil and gas, and other products—the parks are protected for their scenic value. Twenty years ago, as part of an effort to keep the parks in the most natural condition possible, the National Park Service adopted a natural-fire policy that, by letting fire take its course, was supposed to ensure that the parks would be managed more by nature than by humankind.

No one examining the fire and its attendant weather conditions would have guessed that letting the Shoshone Fire burn naturally would in any way harm the park. The natural-fire policy was based on data that suggested that large fires were restricted to old stands of lodgepole pine or spruce-fir. Moreover, Yellowstone's 2 million acres seemed enough to dwarf the damage any fire could do. Anyway, fire was a natural part of the Yellowstone ecosystem. Geological evidence suggests that large fires—probably in conjunction with droughts—occurred there roughly every 250 years, clearing out mature forests and opening the way for new growth.

Though perhaps no one remarked on it at the time, the last round of big fires had occurred around 1700—well over 250 years ago. And—though of course no one could have predicted it—the summer of 1988 would be drought stricken. The time and the conditions for a massive burn would soon collide.

The winter preceding that summer had been dry, but every summer since 1982 had been unusually wet. Park officials, after examining long-term weather forecasts, concluded that the summer of 1988 would be, too. Earlier that year, four fires that had started in Yellowstone—caused largely by lightning strikes—had indeed been snuffed out by rain. During the previous 12 years an average of 15 fires yearly

Flames from a lightning-caused wildfire light up the night in Custer National Forest's Absaroka-Beartooth Wilderness in Montana, just northeast of Yellowstone National Park.

had lit up and burned out, rarely covering more than one acre apiece. The Shoshone Fire, presumably, would die young.

However, the rains failed in July. Yellowstone was drying out. By July 15, nine more fires had started, consuming 8,600 acres. A week later, the affected acreage had doubled and the fires were still burning. That was the day—July 21—that park officials declared all fires still burning to be wildfires and decided to fight them. Three weeks later, Shoshone had combined with another fire a few miles to the south and had seared some 55,000 acres. At the same time, a dozen other fires were at work. Only two were larger than Shoshone proved to be—the North Fork Fire covered at least 69,000 acres, and the combined Clover and Mist fires covered nearly 110,000 acres.

Though in the end more than $150 million would be spent fighting fires in the Yellowstone ecosystem, an 11.7-million-acre area encompassing Yellowstone and Grand Teton national parks as well as parts of six national forests, the conflagration remained virtually beyond control of the 9,000 firefighters who struggled to contain them. The weather was too dry and too hot and, worse still, winds generated by the fire and periodic cold fronts were reaching velocities of up to 100 miles per hour, rushing the fire across the park. August brought Yellowstone the driest weather in a century, as well as a series of cold fronts that powered high winds but did little or nothing to reduce temperatures, which were constantly in the nineties. Moisture levels in dead logs and branches dropped to 7 percent, only a half to a third of normal levels, providing drier fuel for the flames. A few of the seven biggest fires advanced up to 12 miles a day. Impossible to stop, they did not meet their match until autumn rains started in September. Snow finally snuffed them out in November. By then a million acres in the Yellowstone ecosystem—three-quarters of it in the park—had been affected by the fires.

Still, the statistics do not tell the whole story. Park damage was not as bad as the figures suggest—no more than half the affected area had actually burned. Although large stretches of charred forest greet visitors who enter the park—tall, blackened pine skeletons standing rigidly above a sooty, baked earth—the burned areas are in fact patchy. If seen from the air, they look like holes burned into a carpet that is otherwise intact. They do not span vast, unbroken acreages. In total, less than a tenth of the Yellowstone ecosystem was affected.

The embers had scarcely ceased to smoke before biologists and fire ecologists were telling us that, despite the look of things on the ground, the fire had been good for the park. Not everyone agreed with the experts, and, as might be expected, those poorly qualified to understand the biological ramifications of the fires were those most vocal in protesting the way in which the fires had been handled. An inn owner, apparently forgetting that park management is designed to protect the natural working of the ecosystem, not the business interests of local residents, said, "The fire has just literally shut the business down in Cooke City, Montana. We normally run full. We're running nine rooms a night. There's just no business for this time of year." Another resident, with no data to support the claim, said, "Our birds are being lost, we're not having any birds. They're flying away, being

M ilitary transports bring Nomex-clad armies to the 1988 Yellowstone blazes. Hundreds of spot fires, fanned by high winds and fueled by a forest ripe for fire, outlasted and outmaneuvered some 9,000 fire fighters.

burned. We're losing everything." In fact, birds were little affected by the fire.

Predictably, politicians got into the act, responding not to scientific evidence but to voter opinion. With a stunning disregard for facts, Wyoming senator Alan Simpson harangued fellow legislators with this declaration: "Let me tell you, colleagues, the ground is sterilized. It is blackened to the very depths of any root system within." Meanwhile, back in the real world, plants had begun regenerating in most burned areas almost immediately.

Other politicians joined the fray, presumably because 1988 was an election year and local voters were irate about the fire. Montana representative Ron Marlenee, vice-chairman of the House subcommittee that oversees the park service and a long-time opponent of such conservation programs as reintroduction of gray wolves to Yellowstone, said, "I believe that the policy of wilderness lands, of 'lock it up and let it burn,' is in fact a policy of ruin and ashes." Senator Orrin G. Hatch of Utah said the cause of the park fires could be traced to "too much extremism in the environmental area." A Wyoming state representative, Carroll Miller, was particularly lurid: "Now we're told that great benefits are there. Well, I guess then that the bombing of the poor districts of London and Berlin were beneficial because those areas needed to be torn down anyway."

Soon, President Reagan was involved. Avowing that he had never heard of a natural-fire policy, he sent Interior Secretary Donald Hodel to Yellowstone for a major media event at which Hodel, formerly a supporter of the natural-fire policy, suddenly reversed himself and called the fires "devastating" and "a disaster." Eventually a federal task force was appointed to reexamine the natural-fire policy.

Fireweed sprouts like a phoenix from the ashes, two years after the Yellowstone fires. While politicians pronounced the park desolated and the ground sterile, Nature continued her cycle, turning the human catastrophe into fertilizer and a new phase of forest growth. Within two years, most burned areas were covered with seedlings, grasses, and fireweed.

Thus, despite the opinions of various ecological experts to the contrary, Yellowstone National Park came to be viewed as a disaster area, and the entire issue of wildfire—to burn or not to burn—began to rage with all the intensity of a forest blaze itself. Much of the negative feeling about the fire was based on a misunderstanding of the history and role of fire in the natural world, a misunderstanding that could lead to misguided changes in the management and protection of our public lands.

Wildfire: A Burning Need

Thousands of forest and grassland fires break out each year. Many are caused by people, and many by natural forces such as lightning. Even some 20,000 or 30,000 years ago, before humankind ever set foot on

North America, the continent doubtless was singed every year by blaze after blaze. Most would have burned only a few acres. In 1988, an unusually bad year for fire because of extensive drought, more than 70,000 fires broke out, covering more than 5 million acres. That is scarcely 70 acres per fire on average, though clearly some fires covered much more ground and others much less.

It is quite likely that when the first people arrived in North America they had long experience in using fire as a tool. We can guess that they did, at any rate, because their descendants, the Native American peoples, used fire for a variety of purposes. On the plains they set circular fires to force wildlife into the center, where game could be slaughtered easily. In the eastern forests they set fires to clear away trees, creating openings that filled with new plant growth and attracted various herbivores—elk and deer, for example—that could be hunted for food. Possibly the bison of the eastern woods, now extinct, were descended from plains animals that had forged into the woodlands, attracted to clearings created by people.

Native Americans also used fire in warfare. They would set blazes to burn out enemy villages. When Europeans started to explore New World prairies, they found that Native Americans would torch the grasses to deprive the intruders of food for their horses. European settlement itself was characterized by widespread burning, a practice the newcomers learned from Native Americans. Thousands of acres of forest in the East and Southeast were torched simply to clear them for agriculture.

Perhaps because fire was used during settlement to clear land by destroying forest, forest fires at the turn of the century were seen as purely destructive. This was a time when the United States was first awakening to a new conservation consciousness. President Theodore Roosevelt was actively promoting the protection of wild creatures and wild places, and his associate, Gifford Pinchot, was expressing an attitude about fire that had a sort of puritanical moralism about it. "There is no doubt that forest fires encourage a spirit of lawlessness and a disregard for property rights," he said.

Proto-conservationists such as Pinchot clearly thought of fire as a calamity caused by people, or perhaps at best by nature gone awry. It never occurred to them that fire might have a natural role in wild ecosystems, that some plants might not even survive without an occasional fire.

The belief that all fires should be fought became almost indelibly fixed in the popular and official psyche in 1910. In April of that year drought struck the northern Rocky Mountains and the northern plains. Crops failed, and the railroads, suddenly short of freight, laid off thousands of workers. In June, fires broke out across the region. Some were set by the unemployed railroad workers who hoped to be hired to fight the fires. Some were started along rail lines by sparks from locomotives. But since these fires tended to be in areas into

Crow Indians set fire to their enemy's land in a 1905 Charles Russell painting, *Crow Burning the Blackfoot Range.* People have used fire for hunting, warring, and controlling tree growth for centuries.

which fire-fighting equipment could be easily moved, they were usually put out before they could do much damage. The most destructive fires were the 15 percent that were caused by lightning, often in remote, virtually inaccessible areas. Eventually—in a foreshadowing of the Yellowstone fires in 1988—some 10,000 people were enlisted in a fire fight that spanned the northern states from Oregon to Minnesota. About a hundred of those people would die in that fight. When autumn rains and snows finally brought the fires to an end, some 5 million acres had been charred, including 3 million in Idaho and Montana, a million in Oregon and Washington, and the rest scattered across the Black Hills of South Dakota, the Sand Hills of Nebraska, and parts of Minnesota. Fires in Glacier National Park in northwest Montana seared 60,000 acres.

After 1910, Congress gave the Forest Service a broad mandate to create a national fire program. For the next two decades the service experimented with various techniques for fighting fires, including the promotion of livestock grazing on remote national forest lands so that livestock caretakers would be in a position to fight wilderness fires. In the 1930s, fire fighting received a boost from the Civilian Conservation Corps, which put thousands of unemployed young people on the public payroll to fight fires. At this time the Forest Service developed its "10 A.M." policy, which required that whenever a fire broke out, plans would be made to have it out by 10 the next morning. If the first plan failed to make the deadline, another plan would be made to snuff out the flames by 10 the *next* morning, and so on. This became the standard for fire fighting for the next half century.

It eventually became clear that fighting fires was a nearly hopeless task. There were always too many fires and too few fighters. The Forest Service therefore decided to redouble its attack on what it saw as the source of the problem: people. It initiated a public relations campaign to indoctrinate people with the idea that *only they could prevent forest fires.* Actually, early posters have a somewhat more accusatory tone, emblazoning over a pic-

---

WILDFIRE

THE ISSUE Wildfire is a natural part of the life cycle of many plant communities. Without occasional fires, some tree species will not grow. However, fire conflicts with human ideas about wilderness preservation as well as defense of homes built in fire-prone areas, so fires are often suppressed. This can lead to unnaturally intense fires as tinder in the form of dead plant matter builds up in grasslands and forests. Experts continue to debate whether fires should be entirely suppressed or allowed to burn, or whether controlled burns—fires set deliberately to burn off tinder—should be regularly prescribed.

THE CAUSE Fires are caused by natural events, such as lightning, and by human plan or error. Thousands of fires occur in the United States every year, but virtually all quickly die out.

EFFECTS ON WILDLIFE Fire benefits many wildlife species by creating sites for new plant growth, which can be a nutritional boon for herbivores. Many plant species are adapted to surviving fire. For example, some pine species have very thick bark that protects them from burning. The seeds of some tree species will not grow unless fire burns off the outer covering.

EFFECTS ON HUMANS Loss of homes built in fire-prone areas, such as California chaparral country. In all fires there is the risk of losing lives as well. Subjectively, fires may destroy what we see as the aesthetic appeal of wilderness.

EFFECTS ON HABITAT Fire changes habitat. It may clear away trees and permit the growth of grasses and shrubbery. It may keep trees from encroaching on grasslands. It may help maintain a habitat—for example, some southern pine woods are gradually overtaken by hardwood trees if fire is suppressed.

ture of burning forest the words "It's your forest, it's your fault." During World War II, prevention of forest fires took on the trappings of a national security issue. Causing forest fires, the posters suggested, was providing aid to the enemy, which was depicted as a grimacing Japanese soldier.

Two events early in the 1940s brought the need to stop forest fires firmly to the forefront of American thought and, it may be said, made it a moral issue as well. One was the release of the Disney feature cartoon *Bambi*. The climactic scenes in which the forest burns and Bambi and his father, the Old Stag, nearly lose their lives entrenched an antifire concern in the public mind. Predictably, Bambi turned up on early Forest Service posters, but problems with licensing soon forced the service to create its own mascot. After toying with the idea of using a squirrel and even a monkey, the service settled on a bear. Artist Albert Staehle provided the first illustrations of the bear that was dubbed "Smokey." Clad in a ranger uniform, Smokey first appeared on a poster in 1945. Two years later Smokey was given his deathless slogan, "Only you can prevent forest fires," which played on radio with Smokey's voice provided by a Washington, D.C., disk jockey who achieved a bearish basso by speaking into an empty wastebasket. The Smokey we know today was created in 1948, when a Forest Service artist named Rudy Wendelin took over the poster assignment. Smokey himself literally came to life two years later when a bear cub was found wandering Lincoln National Forest in New Mexico after a forest fire. Flown to the National Zoo in Washington, this animal became Smokey the Bear. Twenty-one years later, the aging Smokey was replaced by another cub from the same forest. The original Smokey died in 1977.

Smokey the Bear was a major public relations success. Peaked ranger's hats like the one he wore even came to be called "Smokey the Bear" hats. Bambi and the Bear helped align the public behind the anti-forest-fire crusade.

Meanwhile, the war against fire reached new heights on the technological front. In the 1950s, researchers developed new chemicals with which to suffocate fires. In that very consciously postatomic era, they also studied the hypothetical fires that might erupt during a nuclear war. On the firelines themselves, fire fighters supplemented shovels and axes with modern bulldozers to scrape swaths of earth clean. These zones of bared earth, called *firebreaks*, were designed to stop a fire by starving it. In rugged terrain, fire fighters relied on new aircraft, such as helicopters, to dump chemicals and water on remote flames. Aircraft also transported elite paratrooper squads trained to stop backcountry fires before they spread. Nevertheless, the nation's wartime sense of urgency slowly ebbed, and obsession with staving off wildfires gradually diminished. In 1978 the 10 A.M. policy was dropped.

By then, researchers were learning a good deal about fires and fire ecology. They noticed that fighting small fires often allowed dead wood to accumulate, eventually providing fuel for bigger fires that could not be controlled. Moreover, a new attitude toward natural areas was developing. It stemmed from a report issued by a special federal commission created to study wildlife management in national

▶

A line crew follows the steep trail to a wildfire in Custer National Forest in a portion of the Absaroka-Beartooth Wilderness, Montana. The politics of fire suppression in the United States have evolved significantly since Gifford Pinchot, Theodore Roosevelt's forestry advisor, first declared war on wildfire.

A lightning-caused fire in Custer National Forest, Montana, fills the sky with smoke, while Smokey the Bear reminds visitors to do their part to prevent fires. Smokey is both a national symbol and the only bear with his own zip code. He will celebrate his fiftieth birthday in 1995.

▶

A helicopter drops river water on a blaze in Yosemite National Park. In some remote areas, planes drop bright red retardant, a heavy fertilizer, on fires. In other areas, planes drop paratrooper fire fighters, called smokejumpers.

parks. Headed by wildlife expert A. Starker Leopold, after whom it was named, the committee concluded that the national parks should "be maintained, or where necessary re-created, as nearly as possible in the condition that prevailed when the area was first visited by the white man." This led park officials to test a new policy that encouraged controlled burning and that initiated programs under which lightning fires in remote areas could be allowed to burn. When experience proved that lightning fires rarely affected more than a few acres, the policy was extended in 1976 to nonwilderness areas within national parks. In Yellowstone, for example, fires would be suppressed only in areas where people lived and worked. Human-caused fires or life-threatening natural fires would also be stopped. Other fires would be monitored but left alone. From 1976 to 1987 some 235 lightning fires burned untouched, and unnoted—their effects were too slight to attract the attention of the press.

The long-term harmlessness of natural fires comes as no surprise to

biologists. Fire has long been a part of many ecosystems, and as with any natural force with which living things must contend, plants and animals in those ecosystems have become adapted to fire. Fire is even beneficial to their survival. In dry prairie regions such as the Great Plains, dead plant material accumulates on the ground. Lack of moisture makes decomposition slow, so nutrients are locked up in the dead material, which also shades the ground and may prevent seeds from germinating. If the dead matter were left forever to rot, it would eventually choke the life out of the prairie. But fires, which probably visit any given stretch of Great Plains prairie at least every dozen years or so, burn off the dead litter, providing nutrients for the soil and allowing sunlight to reach the ground and spark the growth of new grasses. This benefits prairie wildlife by providing massive infusions of new food plants. The fires themselves pose relatively little threat to most prairie animals. Birds and large mammals move out of the way. Smaller animals generally burrow underground when fire passes through. Some prairie trees, such as the ponderosa pine, are adapted to fire. Thick, fire-resistant bark protects the trees' inner core.

Southeastern longleaf pine forests are fire dependent. The Southeast has one of the highest rates of lightning strikes on Earth, igniting longleaf pine forests about every three years. The fires burn low to the ground, fueled by fallen pine needles and wiregrass, a bunchgrass species that almost always grows beneath longleaf pines. The pines and the grass depend on fire and together beget a biological conspiracy to spread fires that burn back such competitors as oaks and other hardwoods that would choke out the longleaf pine.

Adult longleaf pines easily survive forest-floor fires. Bark around their trunks is thick and protective, and the more vulnerable branches are high above the ground. Young trees in their first year may not live through a fire, however, though they have some adaptations that give them a chance. The growth bud at the center of the young tree's crown is shielded by mature needles and wrapped in a sheath of younger needles that have a high moisture content. If the flames stay low, the needles provide all the protection needed. Within a few years the young pines grow above the flames, putting their most sensitive and flammable tissues out of danger.

Without occasional fires longleaf pine forests would be overgrown by various faster-growing but less fire-resistant trees, such as deciduous hardwood species. In areas where fires have been suppressed for years, longleaf pine forests were invaded and eventually replaced by loblolly and slash pines and mixed hardwood species. Burning also protects young longleaf pines from brown-spot disease by killing off the disease organism. Though a fire may cause a young longleaf to lose a year's growth, without fire the trees will likely die of brown spot.

The pine barrens that once covered nearly 2 million acres across the Northeast are similarly fire dependent. Without fire, the burn-resistant pitch pines that dominate the barrens succumb to the inroads of deciduous forest. Less than a million acres remain today, and 850,000 of those lie in only two places—750,000 in southern New Jersey and 100,000 on Long Island, New York. Another 100,000 are scattered throughout Massachusetts, Rhode Island, New York, New Jersey,

A golden-mantled squirrel eats dandelion buds in Yellowstone National Park. Burrowing animals survive fire by simply going underground.

▶

A black bear feeds on plentiful plants and berries in Yellowstone two years after the 1988 summer of fire.

▶

Pennsylvania, New Hampshire, and Maine. The barrens and the species dependent on them—such as the endangered buck moth, which feeds exclusively on the leaves of dwarf oak species found in the pine barrens—have been dwindling since European settlement began, destroyed not only by land development but also by fire suppression. Where fire has been long suppressed in the barrens, deciduous trees have taken over. Pristine barrens burned at least every 20 years. All of the surviving portions have burned within the past 50 years.

To Burn or Not to Burn

In some cases the natural fire regime that sustains an ecosystem directly challenges human survival. This conflict underlies some of the most senseless and hard-fought battles between humankind and nature. The classic case can be seen on television almost every year—the fires that plague California's arid chaparral country, destroying houses along with vegetation.

Chaparral is a dense complex of shrubs and trees that covers steep mountainsides throughout California's arid regions. Chaparral cloaks the slopes that hug Los Angeles, keeping erosion in check and filtering water as it flows into streams.

Because lightning is infrequent in the region, fires were probably rare in chaparral before humans arrived. Today the chaparral country is often alight, swept by blazes set either by accident or by arsonists— about 95 percent of the fires are caused by people. When the chaparral goes up, houses burn down.

The combination of houses and chaparral is a dangerous one, says Los Angeles County Fire Department captain Scott Franklin, citing statistics to back up his claim. One in 300 U.S. citizens lives in the Los Angeles area, surrounded by chaparral. Statewide, some 2.5 million buildings and 7 million people are located in high-fire areas. Each pound of chaparral vegetation is the equivalent in energy of a cup of gasoline, and any given acre in the ecosystem bears as much as 60 tons of fuel. If the winds are strong—as they are every autumn when the Santa Ana winds begin—a chaparral fire can move so fast that it is impossible to escape it even in a car. In June 1990, an arson-caused blaze near Santa Barbara consumed 500 buildings. A fire near Malibu covered 25,000 acres. In roughly the first two weeks of August 1990, some 400 square miles of chaparral burned statewide.

Blaming arsonists or accidents or lightning is just a "cop-out" says Franklin. Living in chaparral country, he says, is tantamount to living in a room full of excelsior or shredded paper or with open cups of gasoline—it is just a matter of time before your residence goes up in flames. The danger is multiplied when this normally arid region encounters a drought, such as the one that has plagued California for much of the past decade. Broad-leaved chaparral plants die off, Franklin says, and even oaks begin to die. There is now so much dead fuel from Santa Barbara to Mexico that the region is a powder keg. Should it rain, Franklin says, grass will sprout and, when it dries, provide even more kindling for future fires. The grim realities on which Franklin's concerns are based materialized a year after he made his comments, when the most destructive fire in California history swept through Oakland in October 1991, destroying 2,700 residences and killing at least 19 people. Total cost for the fire was estimated at $5 billion.

The intensity of the California fires is a product not only of the ecosystem. Human ignorance of nature's workings is also a significant factor. Fire suppression has allowed dead material to build up, increasing the intensity of fires when they occur. "Unequivocally, the reason we're having these fires is that the absence of fire has allowed a heavy amount of fuel to accumulate," says Franklin. "There is a gross failure on the part of public agencies to address that problem." He says that dead matter is about twice as dense now as it was prior to a string of serious fires from 1958 to 1962. An estimated 2.6 billion board feet of timber—enough to build nearly 220,000 houses—lies in California's eastern mountain range, the Sierra Nevada. Administrators, he says, provide "loads" of money for fire suppression but nothing for managing fire conditions, such as establishing a policy for the use of regular prescribed burns that would eliminate deadwood before it accumulates to the danger point. "The fire code is a political document," says Franklin. Adds Franklin's colleague, Captain Don Pierpont, "It's designed to provide as much safety as possible with as little financial impact as possible on developers and residents." Because it cannot cover all cases, he says, it is broadly written and therefore vague.

The problem is compounded by inadequate people management. Residents plant various ornamental shrubs and trees that were never meant for life in an arid environment. Because these plants are located next to houses, they provide hazardous fuel during fires. Pierpont says that trees such as the aleppo pine, from Spain, can be quite dangerous. In hot, dry weather pine oils evaporate from the needles of living pines and can easily burst into flames if a fire draws near. The flaming of ornamental plants accounts for one in every five buildings that burn down, Pierpont says. Franklin says the fire department is working to persuade residents to plant certain fire-resistant species, such as native oaks. Many native chaparral species are so fire resistant that they will burn only when dead and dry.

Native chaparral plant species are finely attuned to fire. Many species can resprout from their roots if above-ground parts are burned off. Fire even enhances the growth of many chaparral plant species that produce seeds whose germination is stimulated by heat. The seeds of some other species are covered by a heavy coating that must be burned off for the seeds to open. Moreover, the seeds of some annual plants will not germinate until fire has destroyed certain soil chemicals produced by microorganisms.

Evolution has even caused some chaparral species to enhance the spread of fire. Many plants contain high concentrations of oils and terpenes, such as turpentine, that make them especially flammable, thus encouraging the flames that will burn off old plants and set the stage for new. Chaparral plants are usually shrubby, with many branches rather than a single main stem as a tree would have. As branches age and die, they remain with the living material. In any given chaparral acreage that has not been burned for 30 years, half the plant material may be dead, providing fuel for fire.

Chaparral animal species, too, are adapted to cope with fire, burrowing or moving off as flames approach. But there is one animal that has adapted poorly to the chaparral's natural fire cycle, and that is the human species. People who attempt to buck nature's system are fighting a losing battle. Fires, even if arson and accident could be prevented, must eventually strike chaparral country, and the big fires will always be hard to control. Los Angeles County has 10,000 fire fighters, Franklin says, and they put out the easy ones with speed and alacrity. Big fires, on the other hand, rage out of control, generally costing about a million dollars a day for suppression alone.

Although science has shown that fire is essential to the survival of some ecosystems, natural fire patterns have been changed by human influences. The introduction of roads or communities into wildlands, the effects of frequent human-caused fires in some areas, and decades of fire suppression in others complicate decisions about whether to let burn or put out. Stephen Pyne, a history professor at Arizona State University West with 15 years experience as a fire fighter on the North Rim of the Grand Canyon, says, "What has happened in the last 20 years is a restructuring of fire management around the question of fire in wilderness. This not only involved politics and scientific knowledge, but ethics. How do we relate to nature? And what is good and bad about our behavior to nature? And what's good and bad in nature? The result is that we've gotten into a terribly complicated situation for people managing fire. For a while it seemed that all fires were bad, and it was good to put them out. But we've encountered a kind of smoking existentialism at the end of our shovels. Some fires are good, some are bad, some fires are good but become bad in certain circumstances. There are bad fires that move into another area and may be reclassified as good, there are prescribed fires that get out of control and do damage, there are wildfires that we put out and probably should leave to burn."

The debate is seemingly endless, a squabble between let-burn advocates and put-it-out partisans. The recent incident at Yellowstone helped fuel much of the debate. An examination of what has followed in the wake of the fires of 1988 should provide a few clues about what fires do to wild areas and how wildfires should be managed.

Yellowstone Revisited

Within weeks after the last ember died, Yellowstone National Park was snowed under and the fire lookout who first reported the Shoshone Fire was wintering in Mexico. Fire and drought had drastically reduced the winter food supply for elk and bison, and heavy snows covered much of what was left. The animals had either to dig deep or to move out into thermal areas, where underground heat keeps the Earth's surface free of snow. Some animals fed on tree bark.

But starvation, too, is part of a natural cycle. People persist in thinking that wildlife populations should be stable year to year, but nature apparently abhors stability. In 1988, elk and bison herds were swollen from seven years of mild winters and abundant food. After the fires the pendulum was ready to swing, and, in the guise of starvation, it did so—5,000 elk perished. But the loss probably helped the herd as a whole by removing individuals least able to withstand stress. Older and weaker animals dropped from the breeding population, nature's way of ensuring, perhaps, that only the strongest survive to breed.

For some species—the grizzly, the coyote, and other predators and scavengers—the elk and bison deaths were a boon. For grizzlies, finding a scattering of elk carcasses must have been like discovering a wild-world fast-food chain.

Spring revealed additional good news: Ash had fertilized the

An elk feeds in a Yellowstone Park meadow. Many elk died of starvation during the harsh winter after the 1988 fires, partly because their numbers were unusually high after several mild winters.

◄

ground, and hillsides were covered with pinegrass, asters, and a spectacular bloom of wildflowers. Lodgepole seedlings dotted the forest floor, varying from a few dozen to more than 12,000 per acre. It was clear that Yellowstone had not been destroyed, that it had merely entered another phase, a phase familiar to the mountains that look down upon the park, familiar to the ancient spirit of the place, familiar to the skies and the winds, for Yellowstone has repeated this phase many times in the long continuum that ranges from ashes to growth to maturity to old age to fire to ashes and so on. But for millions of people, who see things only in terms of their brief years as witnesses to the Earth, the park may never again fit their memories of it.

"We've conditioned the public, 'we' meaning our society, by an absolute plethora, an explosion of coffee-table books where everything is lovely and everything is organized and there's no chaos and so on," says Yellowstone superintendent Robert Barbee. "And that simply isn't the way the real world works, that's not the way the natural world is. We don't show the little elk in spring—the calves being crunched by grizzly bears—we show them out there with their mothers, and they're happy, and everything is wonderful. We show the forest as all green. . . . If we have an objective of preserving naturalness, then often times that means chaos and disruption as far as the natural world's concerned. Not tranquil. Not always necessarily pleasing to the eye."

As traumatic as the Yellowstone fires may have seemed initially to nearby towns and to visitors, those flames were as natural a part of the ecosystem as any grizzly, swan, or bison. They play an important role, because as the natural world continues to shrink, parks such as Yellowstone are the last bastions of plant and animal diversity, and fire helps to sustain that diversity. Studies show that fire creates new opportunities for a range of plants and animals, helping to preserve a greater diversity of species.

Fireweed, for example, may appear scarce in Yellowstone's unburned forests, but it survives underground as an extensive root system. When fire clears off the forest, the ground is flooded with nutrients, moisture, and sunlight. Fireweed sprouts and becomes the center of a thriving community of insects. Late in summer, caterpillars gorge on fireweed leaves. Within a few weeks they spin cocoons and emerge as butterflies that migrate south, their travels connecting Yellowstone to ecosystems hundreds of miles away.

Fireweed plays host to a remarkable relationship between two very different species. Ants bring together colonies of aphids, hatched from eggs that the ants keep in their nests during winter. The ants use the aphids much as we use cattle—they bring the aphids to fireweed for food and then feed on a sugary substance secreted by the aphids. All insects are vital to the food chain. They support a variety of predators, including birds such as woodpeckers that comb fire-killed trees infested with beetles.

Fire promotes biological diversity by creating a diversity of habitats. Entire hillsides in Yellowstone appear to have gone up in flames, but the fires actually left behind a patchwork of burned and unburned areas, giving the forest a more complex and varied growth pattern and providing a variety of forage plants for herbivores. And what's good

Two years after the Yellowstone fires, park botanist Don Despain is still learning. Even as the 1988 fires were burning, he was mapping out research plots to inventory seed dispersal and plant regeneration.

for herbivores is also good for predators. The entire food web is served by fire.

"The variation that we see and the different recovery rates, if you want to call them recovery, is that before the fire there were a lot of differences from site to site," says Don Despain, Yellowstone's chief botanist. "North-facing slopes had a lot more plant material, there were more rhizomes in the soil. South-facing slopes had less. The Douglas fir forests had more. The lodgepole pine forest had less. And then the fire itself burns across the forest in different ways. So it depends on the temperature and the relative humidity and the wind speed at the time the fire burns across that place. All of these things go together to make a very diverse pattern of response following the fire."

A botanical team headed by Despain has been visiting burned sites to assess the changes that occurred after the fire, comparing the plant species they find to what was recorded in the same spots before the fire. Tiny aspen seedlings are among the team's most important findings. As mature trees, aspens were scarce in the preburn forests and were declining throughout the park. Despain's studies show that the 1988 fire may help less-dominant species, such as aspen. Though ground fires may have killed aspen trees, their roots lived on to regenerate new growth. When aspens are healthy, they send a chemical signal to their roots that suppresses the growth of new shoots. Fire-caused stress stopped the chemical flow, and the roots responded by sending up armies of tiny new trees. One aspen grove, also in response to stress, produced a huge seed crop. Some of the seeds blew into a marsh, where fire had burned out a deep layer of organic matter and formed a shallow lake. The seeds took root when the water evaporated and could give rise to a new grove, helping to reverse the park-wide aspen decline. On the other hand, aspens may be lost to browsing elk and moose, for which the new tree growth is a bonanza.

Unfortunately, despite all the evidence that the Yellowstone fires posed no biological jeopardy, they are still widely viewed as a catastrophe that should be prevented in the future. Natural-fire plans throughout the nation were suspended pending a review that sought to establish whether Yellowstone officials were unprepared for the severe fire conditions or waited too long to call in fire fighters, or whether the natural-fire policy was itself at fault.

"In no sense of the word was policy responsible for all the fires that occurred in Yellowstone," says Barbee. "I mean, that's just ridiculous. Most of these fires came in from the outside and were man caused and had nothing to do with policy, but they caused a review of policy generally and the application of policy."

Adds Pyne, "The Yellowstone fires occupy, I think, a very special place in the history of American fires. . . . For the last 20 years we've been dominated by a particular question, the question of wilderness fire. . . . I would interpret the Yellowstone fires as really closing out that era. What's likely to drive the system next is the issue of the exurban fire, where you've intermixed wildlands and houses and developments. I don't think any of the questions that were raised in the Yellowstone fires were new questions. They're stuff that the fire community has exhaustively examined for roughly 20 years. And in some ways, they're insoluble. They're based on questions of values, of philosophy, how you perceive your relationship to nature. There's no technical solution to them. What I think the Yellowstone fires do is to end that kind of philosophical discussion. That is to say, we can't afford more large fires like this, in this way. We can't let philosophy substitute for field programs."

Pyne takes issue with the conclusions that Yellowstone officials have reached about the park fires. He agrees with Barbee that no policy problem lay behind the fires. The policy was, he says, wise and necessary. The failing lay in not executing the policy. This happened, Pyne says, because the park lacked detailed plans for making its fire policy work. For example, Yellowstone policy called for natural prescribed fires. A prescribed fire is one set purposefully to achieve some management goal, such as burning off dead undergrowth. A prescribed natural burn is one that is started by nature and then is permitted to burn because it will achieve desirable ends. But to do this, a plan is needed that sets the terms under which a natural fire will be used as a prescribed burn. Pyne says that Yellowstone officials never developed the criteria for determining which natural fires would be allowed to burn. They had, in effect, a let-burn policy, a term that is anathema to park personnel because it implies that they simply let all fires burn.

Pyne says the proof that Yellowstone lacked a sound policy is this: All fires that burned in the region in 1988 were controlled except those in the park. They were the only fires that raged out of control. As for Barbee's comment that most of the fires started outside the park, Pyne asked why Yellowstone officials accepted those renegade flames. He concludes, "I think they goofed. But we all goof." No one, he suggested, made sure that Yellowstone had a sound fire policy. The result is not an ecological disaster, Pyne says. The real disaster was the loss of funds used fighting the fires, he says.

The biggest fault that Pyne finds with fire management in Yellowstone is that it ignores the role of human-generated fires during the past few thousand years. He suspects that Native Americans regularly burned over parts of the area, getting rid of incendiary undergrowth and stemming off major conflagrations. When the National Park Service attempts to keep parks in a pre-Columbian condition, it overlooks the presence of Native Americans, as if presuming that there were no human impacts prior to the arrival of Columbus. Failure to set controlled fires in the park, as Native Americans probably did for perhaps the past 10,000 years is, he suggests, as much an interference with "natural" procedures as is fire suppression.

While Pyne and some of his colleagues ponder the role of fire and humanity in Yellowstone, the federal review committee has determined that the National Park Service's natural-fire policy will survive, though like the park itself the policy has evolved as a result of the fires of 1988. Aimed squarely at avoiding another conflagration, a slightly modified policy calls for more intensive monitoring of fire conditions and greater responsiveness to surrounding communities. As James Woods, host of the "Wildfire" Audubon television special, observed at the conclusion of the program, "When and where to let natural fires burn has always been complicated. We know that putting them out is not always the right answer. One lesson to learn is that just as we struggle to protect wilderness areas, so must we also work to preserve natural processes within them, including fire. The question remains: Can we learn to live with nature's fire?"

Perhaps the answer to that question can be found by observing events occurring in the park right now. Nature reveals itself in strange ways, and one revelation in Yellowstone lies in the activities of a bird called the Clark's nutcracker.

Every autumn the nutcracker comes to the white-bark pines that perch on the park's rockier ridges, fulfilling an essential role in pine reproduction. The birds pick apart thousands of white-bark pine cones, collecting up to twenty seeds per cone. Holding the seeds in a pouch just below their beaks, the birds fly off to all sections of the park in search of places to hide the seeds.

Purple fireweed and other wildflowers gradually reclaim the forest floor two years after the Yellowstone fires. Fireweed is so named because it often is the first to sprout after a blaze.

They land in burned areas, or wherever soil is exposed, and bury the seeds in places only they know how to find when spring melts the snows and the time for retrieval arrives. By leaving some seeds unretrieved, the nutcrackers have begun to replant the 60,000 acres of white-bark pines lost in the fires. Ironically, 1988's seed caches will produce the biggest wave of new pines because the fires destroyed the trees the birds use as landmarks for finding buried seeds.

While we debate, nature rebuilds.

WHERE THE DEER AND THE ANTELOPE PLAYED

T he American West is rich in legend, a place marked in history with a fabled beginning—the search of Spanish conquistadors for cities of gold. By the time the West fell into American hands, purchased from Napoleon by Thomas Jefferson, its true nature was still partially obscured by the fog of myth. Jefferson, when he sent Lewis and Clark to explore the new lands, fully expected them to return with the fresh hides of woolly mammoths, little knowing that the long-haired elephants had vanished 5,000 years before.

Even the truth about the American West had the trappings of fable. Wildlife thrived there in unbelievable numbers. The West was home to perhaps 50 million bison and as many pronghorn antelope. Its mountains, valleys, and plains were thick with deer, elk, wild turkey, wild sheep. Throughout the West roamed creatures that travelers from the East had never even imagined, such as the highly aggressive bear we call the grizzly and they called the grisly. Raised on the much smaller and shier black bears of eastern woodlands, early explorers were astonished by the West's king predator, a massive creature that was as likely to fight as to run. Lewis and Clark wrote in amazement about how many bullets they had to fire into one to kill it.

Travelers along the West Coast also encountered a huge flying bird the like of which they had never seen: the California condor, a sort of vulture that was the largest flying bird in North America, second in size in the entire world only to the Andean condor. Its wings spanned nearly 10 feet and the wind whistled as it passed through the feathers at the tips.

The condor and the grizzly would both virtually pass from the West, surviving today only in zoos or within the confines of a handful of national parks. The seemingly endless bison and pronghorn herds would vanish too, slaughtered for market, their great living spectacle transformed into hundreds of thousands of carcasses rotting beneath western skies, reduced finally to bones that melted into the prairie soils. Thus the Old West ebbed away.

Yet some essential part of that fabled place survives today on the far horizon where the West begins. Geographers will tell you that that horizon lies at the hundredth meridian, but their criterion is too narrow. The real West lies in the imagination, in a place where cities and civilization falter and fade away, a place where men and women are just a little more independent and self-reliant and the horizon is just a little farther away and a lot wider, a place where distant mesas rest pale blue against a paler blue sky and the scent of prairie grasses perfumes the wind, a place where the moon and the stars shine brighter and coyotes cry in the evening, a place where eagles soar on outstretched wings in cloudless skies, where open spaces dwarf the tallest mountains, where you can feel the ghosts of Crazy Horse and Sitting Bull, the restless souls of Wild Bill and Cole Younger, and hear the silent thunder of phantom bison rushing across the plains.

◄

A plains buffalo, or bison, gets its feet wet in Yellowstone National Park. Bison once roamed the western ranges in vast numbers. Now only isolated, protected populations are to be found.

For the lucky few who still work the land, the West is a living reality. It lives in early spring mornings when you still feel the chill of winter as you walk from house to barn for the first milking of the day. It lives in calving time when frost still clings to the land at dawn and you never sleep more than two hours straight because you have to check the cows and their newborn calves to be sure all's well at a critical time. It lives in memories of days on horseback, scorching under summer sun, or riding for home in a sudden thunderstorm tense with the knowledge that in this open country you are the surest target for lightning. You feel the West forever in the callused palms of your hands, in the scent of a hayfield newly mown and tractored into stacks, in the remembered warmth of a branding iron smoking hair from the ribs of a calf you helped hold down one bright spring day.

The West is romance pure and simple. What is more romantic than the notion of the lone pioneer eking out a living from a harsh, uncompromising land? The cowboy on horseback speaks to anyone moved by a struggle for survival against overwhelming odds, for the cowboy is pitted against sun, rain, and blizzard, against cyclone and drought, against his own limitations and physical endurance, and he meets that struggle with the stalwart individualism that typifies all our mythic heroes.

What a shame, then, to examine the current state of affairs in much of the American West, where we find government subsidies are buoy-

The vastness of the Southwest's landscape makes the predominant four-footer look minuscule. For all its apparent smallness, the bovine has done disproportionate damage to the land. These cows near Rio Puerco, New Mexico, have helped turn the area into a moonscape.

ing up the cowboy, where the lone figure on horseback is quite likely riding off to get a check from the public dole, where the struggle against nature has become so one-sided that the cowboy and his cows are more or less slowly and more or less rapidly destroying vast parts of the vast land in which they live, and federal tax dollars are helping them do it.

The Birth of a Hero

It was on the second day of January in 1494 that Christopher Columbus landed near Cape Haitien on the Caribbean island of Haiti and unloaded 34 horses and an unrecorded number of cattle, thus bringing into the New World, for the first time, all the basic ingredients needed to create the American cowboy.

Cattle did not reach the North American mainland until 1518, when Gregorio de Villalobos and his men landed on the banks of the Panuco River, near where Tampico, Mexico, would one day take root, and unloaded the continent's first shipment of calves.

The Spaniards favored beef for their tables, so more cattle soon followed, and ranching quickly became a source of wealth in the New World. One of the first ranchers was Hernando Cortez, the religious zealot and lawyer who landed in Mexico in 1519 and subsequently conquered the Aztecs and went in search of a city of gold. When that endeavor proved futile, he divided up among himself and other landowners the Native Americans he had subjugated, establishing a system of New World serfdom, and set up a cattle ranch south of Toluca. In the warmth of the central plateau his herds soon grew larger. To keep track of his stock, he marked his cattle with a distinctive brand of three Latin crosses. Other Spaniards took up ranching, letting their cattle roam free, and soon disputes arose among them as to who owned which cattle. The branding system was supposed to make ownership clear, but some ranchers used the same brands as others. Also, many an enterprising cattleman would capture the unmarked calves of other ranchers and put his own brand on them. Thus two traditions of American ranching appeared early in New World history—the use of hot irons to brand calves and the purloining of strays, which cowboys would later call *mavericking*.

By the end of the first quarter of the sixteenth century, Spanish cattlemen were constantly at each other's throats over cattle ownership and grazing rights. Their Native American neighbors frequently protested that the Spaniards let their free-roaming cattle trample native crops. To keep everyone happy, Mexico City's town council—composed of cattlemen—created a colonial livestock association called the *Mesta,* which quickly set about looking after the ranchers' interests. Thus were established two more traditions of the American West—ranchers as a political force and the application of that force through cattlemen's associations.

The Mesta created the first official registry of brands and ensured that no two brands were alike, thus helping to diminish disputes over who owned which cattle. It also helped settle disputes over grazing

rights, with some help from the heavy hand of the Spanish crown. In 1533 the crown, in an effort to cut a burgeoning number of New World court cases revolving around who had the right to graze cattle where, decreed that all cattle—those of Spaniards and of Native Americans alike—would be grazed on common grounds. At first the ranchers balked at the idea of putting their cattle with those of Native Americans, but the Mesta saw to it that the ranchers complied. This cut the number of court cases, but it also created a new complication: Putting hundreds of cattle in a common place provided the ideal conditions for the development of a new enterprise—cattle rustling. The Mesta cracked down on this, too, imposing fines and meting out punishment.

Nevertheless, as the herds grew, keeping track of them became increasingly difficult. Ultimately, to cut back on losses, the Spanish ranchers concluded that they needed to keep an eye on the cattle at all times. They did not reach this conclusion willingly, however, because they thought guarding cows was beneath their station in life. They soon concluded, though, that it was not beneath the station of the defeated Native Americans, and they quickly put natives on horseback and set them to work watching cattle. Thus, finally, in addition to such things as cattle, cattlemen's associations, branding irons, horses, and rustlers, we have the final ingredient of the cowboy world—the cowboy himself. In the centuries ahead he would become a hero to millions, written about in nineteenth-century dime magazines and later riding roughshod across thousands of movie screens, two-fisted and revolver-armed, recreated in the images of Gary Cooper, John Wayne, Tom Mix, Roy Rogers, and the rest. But in the beginning the cowboy was a sort of slave or feudal serf on horseback trying to keep rustlers from stealing livestock. He had the social status of a cow pie.

Meanwhile, cattle herds were slowly seeping north from the central plateau, following in the path of the meat-hungry mining camps that proliferated when silver was discovered near Zacatecas. As cattle spread throughout Mexico, some rambled off beyond control of their human captors and found refuge in distant reaches where they and their offspring might pass their whole lives without seeing a human. Unlike most modern breeds, these cattle were aggressive and quick to charge, especially if hitherto untouched by human hands. When, starting in the 1550s, the Spaniards decided to round up the feral animals, they found themselves battling against genuinely vicious animals armed with rapier horns backed by nearly half a ton of muscle and the sort of uncompromising disposition that comes from a lifetime of freedom. The Spaniards reasoned that if uncivilized cows and bulls were dangerous, then it was best to keep them from becoming unaccustomed to humanity. With this idea in mind, the ranchers started annual roundups that they called *rodeos*. A rodeo was not a sporting event but a series of hard days or weeks rounding up strays into herds. Such roundups remained a necessary part of ranching for as long as ranchers continued to practice open-range grazing, in which livestock roamed most of the year at will, feeding and watering wherever they chose, like wild animals. This system, too, was precedent setting, for open-range ranching would dominate the cattle industry of the American West for the next three centuries.

Although in the 1500s cattle ranching was already spreading beyond the Mexico City area, during the sixteenth and most of the seventeenth century the Spanish cattle industry stayed south of what would one day be the U.S. border. This was partly a logistical matter. Mexico lacked the people necessary to settle the more northerly lands, and the native peoples there—unlike the Aztecs and other more urbanized natives of Mexico—were hostile and aggressive and much preferred shedding blood to falling under the Spanish yoke.

Eventually, though, politics compelled the Spanish to cross the Rio Grande. In the 1680s France started moving into the area now called Texas—the explorer La Salle founded a colony at Matagorda Bay—and this stimulated the Spaniards to establish a better hold on the New World lands they claimed were theirs. They had some unexpected and unintentional help in this endeavor when some native warriors wiped out the French colony. Their task thus simplified, the Spaniards further bolstered their claim by setting up missions in Texas. The missionaries were supposed to convert the Native Americans to Christianity and, doubtless, to servitude, since one seemed to follow the other. In any event, the first Texas cattle were probably those brought to a mission built on San Pedro Creek north of Weches, Texas, in 1690. A Captain de Leon brought 200 cattle there, but left only a bull and a cow as part of a seed operation. On his trip back to Mexico, the captain left pairs of cattle here and there at various rivers and streams to establish herds. Of course, ensuing expeditions brought more and more cattle, so by the early 1700s Texas was becoming infested with the beasts. By the end of the eighteenth century, some 100,000 cattle inhabited Texas, roughly 35 times as many cattle as colonists. This supply of livestock just met the demand for them, making beef a source of prosperity except in years when the Spanish crown put strict limits on trade. The crown, for example, tended to discourage trade between Texas and French Louisiana, even though Louisiana markets were closer to Texas than were markets in Mexico.

During the seventeenth and eighteenth centuries, cattle also reached other parts of Spain's New World empire, including Arizona, New Mexico, and California. During this time, a new order of things was being established in Spain's North American empire, and it set the stage for a period of rapid change that in the nineteenth century overtook the *vaquero,* or Spanish cowboy. In 1821 Mexico declared itself independent from Spain. Without royal support, the missionaries in Texas fled for home, and the Texas mission system collapsed. This created a void in the region's settlement pattern, and the void was soon filled by land-hungry Americans who rushed in after Mexico approved an 1825 immigration bill opening Texas to American settlement.

The Americans came as farmers. Ranching, with its emphasis on livestock rather than crops, was alien to them. But though they harbored no intention of becoming ranchers, they soon found reason at least to *claim* an interest in livestock breeding. Mexico gave farmers land grants of up to 277 acres. Avowed cattlemen received up to 4,338. Claim to be both, and you could net better than 4,600 acres.

During the era of American settlement, about 80 percent of the 100,000 or so cattle in Texas were feral Spanish breeds. It is not clear exactly what breeds were introduced by the Spanish, but at least one was the Andalusian fighting bull. The other two breeds, one a white animal with black markings and the other a tan to red creature with a narrow head, were brought into Texas for beef. The American colonists set about capturing strays of these gaunt, ill-tempered breeds to build up herds. The cattle were feral and vicious, and settlers generally tried to avoid meeting them on foot. Coming near them even on horseback could invite a goring. Once captured, these animals were crossbred with American cattle breeds, which gave rise to the Texas longhorn. The new blood did nothing to the Spanish descendants' temperament. Longhorns remained vicious and round-ups risky.

Like the longhorns, the American settlers became hybrids of a sort. They learned the cattle business from the Mexican vaqueros and even adopted some of the vaqueros' dress and equipment, such as the heavy saddle with a curved horn at the front, a supportive pommel at the back, and long stirrups, all designed to make the rider more efficient at chasing down cattle and staying mounted while roping them.

It was around this time that the term *cowboy* came into vogue for describing a Texas cattleman. It was a product of the 1830s war in which Texas fought for independence from Mexico. During the war,

Americans in Texas, as part of the war effort, began rustling cattle from Mexican ranches along the border. These rustlers were called cow-boys, as it was then spelled.

In those days, *cow-boy* had slightly pejorative connotations, according to David Dary's entertaining book *Cowboy Culture,* from which much of the information in this section is drawn. The term apparently has an Irish lineage, first popping up about a thousand years ago in reference to horsemen who worked with cattle. In the 1600s, Irish cow-boys who fell into disfavor with the British crown were sometimes given a choice between imprisonment or emigration to the American colonies. Cow-boys who chose the latter option were indentured to colonial farmers, so the term came to connote someone with a shoddy or criminal background. During the Revolutionary War, British Loyalists who rustled cattle from American Patriots were also known as cow-boys, which did nothing to improve the word's image. This heap of unsavory connotations made the term *cow-boy* seem pejorative in the minds of most Texans, so the cattle handlers themselves preferred to be called vaqueros. After the Civil War, as ranching spread throughout the West, cow-boy took on the meaning of "one who works with cattle" but also still tended to connote "one who rustles cattle," or at best "one who is a common laborer." Sometime around 1900 the hyphen fell out, and the word became *cowboy.* Presumably it was not long after that, perhaps because of the motion-picture industry's fascination with the western and the popularity of such western writers as Zane Grey, that *cowboy* took on less the trappings of crassness and more those of heroism. At any rate, by the time Texas won independence from Mexico on April 21, 1836, with the defeat of the Mexican army by General Sam Houston and his troops, *cowboy* was becoming part of the common parlance.

At the time of independence, about 30,000 people lived in Texas, and about 20,000 of them were Americans. By the 1840s, those Americans were adopting the Mexican approach to cattle ranching. Instead of fencing cattle in or keeping them near home, the novice ranchers began letting their animals roam the open range.

In 1845 Texas became the twenty-eighth state, an ambiguous development because Mexico did not accept the Rio Grande as an international boundary and claimed parts of Texas up to the Nueces River. This resulted in a war between the United States and Mexico, and when the war ended in 1848, the United States had won not only Texas down to the Rio Grande but also the region that now encompasses New Mexico, Arizona, and California. Within days after the war was resolved, word arrived that gold had been discovered in California. The rush was on, drawing thousands of prospectors to the West Coast, and cattlemen sensed a profitable opportunity. By 1850, more than 100,000 cattle were being driven to California from Texas, Missouri, and other states. It was the beginning of the era of cattle drives, in which cowboys herded along up to a thousand animals at a time on journeys that might take a full five months. Ranching developed as a major industry in California itself as immigrant Americans supplanted the Spaniards who had raised cattle in California for nearly 200 years. Cattle from southern California were driven up the coast to the northern markets that served the prospectors. But it was the advent of the railroad that gave birth to the golden age of the cattle drive. During the 1850s, cattlemen in Texas began driving stock to markets in Springfield, Missouri, and even as far as Chicago, where they could be put on rails. The Civil War slowed the ranching business in the early 1860s, when Texas fought on the side of the Confederacy. After the war, returning ranchers had to repair neglected houses and round up cattle that had roamed free for nearly five years. But better times were ahead. By 1870 the railroad had cut across Kansas and Nebraska. This triggered the start of the famous cattle drives from Texas ranches to railheads in Dodge City, Wichita, Abilene, Ogallala, and Omaha. From there cattle were taken by train to eastern cities.

Getting the cattle to the railheads was the job of the cowboy, who was now becoming a bit more respectable. He deserved whatever respect he got, for he succeeded at a difficult task, moving perhaps a thousand cattle and several hundred horses on a journey of perhaps a thousand miles. Despite the popular image of the cowboy as the quintessential individualist, cowboys in reality were team workers, and egos were kept in line.

The cowboys who brought livestock north were usually hired hands from Texas who wanted to set up their own operations, but some were from the East, too. Notable among this group was Theodore Roosevelt who, in 1884, bought two ranches in the Dakota Territories in an attempt to escape the grief he felt when both his first wife

Teddy Roosevelt is depicted rounding up the steers (marked as states) in a 1904 political cartoon by Clifford Kennedy in the *Washington Post.* Although he failed as a Dakota Territory rancher, Roosevelt retained colorful memories of his three-year stint as a cowpoke.

and his mother died on the same day. Roosevelt has left colorful accounts of his ranch life, which lasted three years before financial losses caused him to sell out and move back East. In books about his life as a "ranchman," he provided glimpses into the development of the northern plains as cattle country and offered hints of what life was like on a ranch in the 1880s:

> It is but a little over a half a dozen years since these lands were won from the Indians. They were their only remaining great hunting grounds, and toward the end of the last decade all of the Northern plains tribes went on the warpath in a final desperate effort to preserve them. After bloody fighting and protracted campaigns, they were defeated, and the country thrown open to the whites, while the building of the Northern Pacific Railroad gave immigration an immense impetus. . . . Some Eastern men, seeing the extent of the grazing country, brought stock out by the railroad, and the short-horned beasts became almost as plentiful as the wilder looking Southern steers. . . .
>
> In its present form stock raising on the plains is doomed, and can hardly outlast the century. The great free ranches, with their barbarous, picturesque, and curiously fascinating surroundings, mark a primitive stage of existence as surely as do the great tracts of primeval forest and, like the latter, must pass away before the onward march of our people; and we who have felt the charm of the life, and have exulted in its abounding vigor and its bold, restless freedom, will not only regret its passing for our own sakes, but must also feel real sorrow that those who come after us are not to see, as we have seen, what is perhaps the pleasantest, healthiest, and most exciting phase of American existence.

The open range was brought to an end by the settlement of the West. As more and more farmers moved in, cattlemen found themselves increasingly bearing the brunt of complaints that free-roaming cattle were destroying crops. At first the ranchers could ignore these complaints because they held political power. But as the number of farmers grew, so did the farmers' political power, and politicians started listening more closely to their complaints. Soon cattlemen faced laws that required them to keep their cattle off of other people's property.

To do this, ranchers hired range riders to patrol ranch perimeters and herd stray cattle back onto ranch lands. This was fairly impractical, though, so eventually ranchers faced the prospect of fencing their lands, using one of the many varieties of barbed wire that first hit the market in 1874. Sale of barbed wire grew from 5 tons in 1874 to more than 40,000 tons only six years later. Barbed-wire fences, of course, were the death knell of the open range. The ranchers were not happy about it, but there was little they could do but lament the change, as one Texan quoted in Dary's *Cowboy Culture* did in 1884:

> These fellows from Ohio, Indiana, and other northern and western states—the "bone and sinew of the country," as the politicians call them—have made farms, enclosed pastures, and fenced in waterholes until you can't rest; and I say, Damn such bone and sinew! They are the ruin of the country, and have everlastingly, eternally, now and forever, destroyed the best grazing land in the world. . . . I am sick enough to need two doctors,

a druggery, and a mineral spring, when I think of onions and Irish pota-
toes growing where mustang ponies should be exercising, and where four-
year-old steers should be getting ripe for market. Fences, sir, are the curse
of the country!

Fences are indeed quite likely a curse on any fine open place. But
by the turn of the century, any rancher looking for another dreadful
curse might have done well to take a peek in a mirror.

A Curse Upon the Land: Grazing on the Public Domain

Early in the nineteenth century, the public domain consisted of all
lands in the area today encompassed by the lower 48 states, with the
exception of the original 13 colonies and Texas. Eventually, some 1.4
billion acres were under federal administration. Because the federal
government through the years has tried to divest itself of its lands,
today only about half a billion acres remain under federal administration. The other billion or so were sold or given away.

Disposal of the land was handled by the General Land Office, which
was created in 1812 as part of the Treasury Department and was not
transferred to the government's chief land-management agency, the
Department of the Interior, until 1849. At first Congress tried to sell
off the lands to earn revenue for the government. When this failed,
Congress simply started granting lands to the states for such things
such as the construction of railroads, canals, and roads and to support
local schools and colleges. In 1850, Congress began giving land directly to railroad companies as part of a plan to encourage the expan-
sion of the rail system. This piecemeal approach to giving away public
lands created a checkerboard pattern of land ownership that still
makes consistent land management a serious challenge in the West.

Cattle grazing entered the picture after passage of the first homestead act in 1862, which allotted 160 acres to anyone who would live
on the land for five years and develop it. This was an attempt to encourage settlement and farming, but it failed miserably in the arid
West. The land was too dry for farming and the grants were too small
for ranching. It was not until 1916 that Congress finally tried to rem-
edy this problem by authorizing homestead grants specifically for
ranching and boosting the size to 640 acres, which was still too small
to do anyone much good.

The small grants aggravated the checkerboard pattern of ownership.
Homesteaders interested in ranching would, in the early days, sign up
for 160 acres surrounding a water supply. Once water was secured,
the homesteaders would simply let their cattle graze on adjacent public lands. Since the cattlemen did not own the land, which was equally
available to anyone else who had established a source of water, they
were tempted to put as many cattle on the range as possible in the
hope of making a big profit. Because they did not own the land, they
did not care if it was harmed by overgrazing. This abuse of public
lands led to overexpansion of the cattle industry during times of peri-
odic heavy rains and resulted in greater cattle densities than the land
could sustain once dry periods returned. Fencing of the range only

compounded the problem by confining cattle to relatively small areas.

The late nineteenth century saw more and more cattle and sheep being piled onto western public lands. Overgrazing—the destruction of grassland by more livestock than the land can feed—was reported as early as the 1870s. Combined with bad weather in the 1880s, overgrazing led to disastrous losses of livestock. By the turn of the century, even ranchers, who traditionally opposed any sort of government intervention, were calling for federal help in establishing range-management plans that would reduce grazing excesses and conflicts about grazing rights. Nevertheless, during the first three decades of the twentieth century, all attempts by Congress to regulate grazing were defeated by states' rights activists and by politicians committed to homesteading. The only sign of progress occurred in 1906, when the Forest Service began to charge grazing fees on its lands and simultaneously reduced grazing levels. However, this had a boomerang effect, because the changes hardened rancher opposition to all other attempts to control grazing.

In the 1930s two factors contributed to the passage of the first range-protection law. First, a Montana grazing experiment, in which a local stockmen's association controlled use of a grazing site under the direction of the Department of the Interior, showed that grass production doubled within three years after well-regulated grazing was initiated. As a consequence, the number of cattle that could be grazed at the site also doubled. Congress was impressed.

GRAZING ON PUBLIC LANDS

THE ISSUE Federal policy dating to the nineteenth century permits ranchers to graze livestock on western public lands. Critics believe that the fees charged the ranchers are much too low—less than a fifth the usual fee for leasing private lands—and that ranchers put more livestock on the range than the land can sustain. Vast regions of the West have been denuded or degraded by livestock, causing increased erosion and disruptions in stream flow. Much of the federal public land is in poor condition because of overgrazing.

THE CAUSE Because the public land is not owned by the ranchers, they feel compelled to put as much livestock on the range as they can. Livestock then eat more vegetation than the grasslands can replace in most years, causing degradation of grazed areas. Also, cattle tend to congregate near water, causing trampling of streamside vegetation and increased erosion of stream banks as well as muddying of streams. Siltation chokes out stream plants and animals.

EFFECTS ON WILDLIFE Wild species are forced to compete with livestock for food and water. Some fish species have become much reduced or extinct because of the loss of stream habitat. Similarly, the destruction of streams has led to a loss of riparian vegetation and a decline in nesting places for many birds.

EFFECTS ON HUMANS If grazing continues, the habitat will become so degraded that livestock will not be able to use it. This could have local economic effects in the years ahead. The presence of livestock and the degradation of surroundings are a problem for recreational industries associated with such activities as hiking, camping, and fishing. Public-lands ranchers face a potential loss of immediate income if the public lands are closed to them or if they are compelled to reduce grazing to protect the land, though in the long run restoring the habitat is a potential benefit to livestockmen as well as to wildlife and wildlife enthusiasts.

EFFECTS ON HABITAT As the most nutritious and desirable food-plant species are consumed, they are replaced by less palatable plants. Often, livestock can survive on this less nutritious vegetation because of supplemental feeding by ranchers, but for wildlife the range loses much of its value. The loss of grassland cover also leads to heavy erosion. Streams become clouded with mud, and the darker water warms under the sun. Consequently, fish and other species that need clear, cool water die off. Trampling around streams leads to destruction of streamside vegetation. Young trees are eaten when soft and succulent, so old streamside trees are never replaced. Eventually, the trees die off, causing widespread loss of nesting and roosting habitat for birds and other animals that use streamside woodlands.

The second factor was a force of nature. In 1934 the nation suffered the worst drought and dust storms in its history. As the Dust Bowl years began, land eroded by sodbusting and overgrazing disappeared on the wind. The surface soils of Oklahoma, Texas, and other parts of the West ended up in New York City and Washington, D.C.

That summer Congress responded by passing the Taylor Grazing Act, which was intended to end grazing excesses and land damage by initiating a permit program that would control the number of cattle on the range and exact a fee from permittees. Revenue from the fees would go to range improvement, such as water wells and fences.

The law was to be administered by the new Division of Grazing, renamed the Grazing Service in 1939. The service's first chief was a Colorado rancher, Farrington Carpenter, and he set up grazing advisory boards that were supposed to cooperate with federal agents in developing grazing plans. Carpenter gave the boards a strong voice in determining the direction of management plans and then saw to it that a 1939 congressional amendment to the grazing act required that each board be composed of five to twelve stockmen and one wildlife specialist.

Thus was the deck stacked against federal agents' seeking to control overgrazing and raise grazing fees. With the aid of congressional allies, ranchers were able to block changes in range management. This made it impossible for the Grazing Service to raise the funds it needed to bankroll its operation. Then, when other congressional members—especially those on the House Appropriations Committee—noticed that the Grazing Service was not collecting sufficient grazing fees to cover its administrative costs, they concluded the agency was faulty and drastically cut its budget and personnel. To survive, the service had to be supported by funds from the rancher-dominated grazing-district advisory boards. The original purpose of the Grazing Service was therefore all but discarded.

President Harry Truman tried to correct this situation in 1946 by consolidating the General Land Office and the Grazing Service into one new agency, the Bureau of Land Management. Though the bureau set out in the 1950s to address some concerns about wildlife management and natural-resource conservation and also began to attract significant numbers of range-management professionals, progress was slow until the 1960s, when Congress enacted laws requiring the bureau to broaden its management goals to include wildlife protection.

Despite the laws, the restructured bureau, and a better scientific understanding of grassland biology, shockingly little has changed on public rangelands since 1900. The Bureau of Land Management still tends to kowtow to the whims of western ranchers, about 33 percent more cattle are grazed on public lands than the land can support with any semblance of ecological health, grazing fees are still pitifully low and result in a net loss of federal revenue every year, and more than half of western public range lands are in only fair to poor condition, which means that they are, by all biological standards, ruined ecosystems.

▶

A hawk, probably a Swainson's, leaves its perch to nail a mouse or other tasty morsel. Many ranchers view raptors and scores of other animals as cattle-country pests that kill livestock. Ranchers also are the main force preventing reintroduction of wolves into Yellowstone National Park and on public land in the Southwest.

Death in the West, or, Eaten Alive!

Grassland areas on public lands scattered throughout the West are dying, and it is the livestock industry that is killing them. About 220 million acres of the 260 million open to cattle and sheep grazing are managed by the U.S. Forest Service and the Bureau of Land Management. These lands are leased to livestockmen for a fee of $1.97 per month for every cow or every five sheep. Under this system, some 7.5 million cattle and nearly 3 million sheep are put on public lands each year by roughly 23,000 permit holders spread over 16 western states. The ranchers who graze livestock on the public lands for at least part of each year account for only about 5 percent of all American stockmen. They are getting a big break. For one thing, the fees they pay are only a fourth to an eighth the going rate for leasing private land. This is doubtless a great help to the small rancher, but according to a report by the House Committee on Government Operations, less than 10 percent of federal forage reaches stock belonging to small operators. The other 90 percent—and the cheap fees—goes to corporations such as Getty Oil and Union Oil and to wealthy individuals, such as J. P. Simplot, who is both the largest grazing permittee in Idaho and reportedly the state's richest citizen. Nearly a million acres of federal land in Oregon is leased to the Vail Ski Corporation.

The cheap fees amount to a rancher subsidy. Presuming—simply for the sake of comparison—that monthly fees on private land were set at $5.20 per head (scarcely more than double the federal fee), the federal charge of $1.97 represents an annual subsidy of about $60 million. That amounts to roughly $2,600 per permit holder (of which there are 23,000) and, since grazing leases are for 10 years, adds up to $26,000 over the duration of the permit.

But that is not the only loss to the federal treasury. A report by the General Accounting Office concluded that the Bureau of Land Management's grazing program—which includes development of watering sites and planting of grass (including exotic species) for livestock—cost taxpayers $39 million in 1986, a year in which grazing fees amounted to $14.6 million. That comes to about $1,100 per year for each of the permit holders if you subtract the grazing fees from the $39 million in costs. That adds up to $11,000 over the duration of the permit. Added together, the two subsidies come to $37,000 per rancher in the course of a decade.

The grazing lessees bring other costs to the taxpayer, too. One of these is the federal Animal Damage Control Program, run by the Department of Agriculture. The annual budget is about $25 million—nearly as much as the federal government spends on endangered-species protection each year. Primary targets of interest to ranchers are coyotes, mountain lions, black bears, and prairie dogs. In fiscal year 1988, federal agents killed 76,000 coyotes, 203 mountain lions, 9,127 black bears, and 124,292 prairie dogs. In past decades control officers working at the behest of ranchers helped wipe out the gray wolf and grizzly bear, both of which are now protected by the Endangered Species Act and are subject to programs designed to help revive them. Presumably if the agents could succeed in nearly wiping out coyotes and prairie dogs, these animals too would be added to the endangered

▶

Each year the federal Animal Damage Control program, operated by the U.S. Department of Agriculture, poisons, shoots, and traps thousands of coyotes and black bears, hundreds of mountain lions, and tens of thousands of prairie dogs. This taxpayer-funded war on wildlife adds up to an additional rancher subsidy.

species list so the government could try to recover them. To help ranchers, who claim livestock losses to predators and loss of grass-lands to prairie dogs, federal agents kill badgers (which feed on prairie dogs) and coyotes (which also feed on prairie dogs) in each of the eleven western states in which they occur, bobcats in every western state except Washington and Arizona, mountain lions in every west-ern state except Wyoming, and prairie dogs in New Mexico as well as in three states that lie outside the eleven western states—Texas, Okla-homa, and Nebraska.

The ranchers also receive help from various Department of Agricul-ture assistance programs, including the emergency feed program that subsidizes their feed expenses during droughts. Taxpayers spent $140 million on emergency feed in 1988 to help ranchers keep livestock on devastated land too arid for grazing in the first place. As Stanford bi-ologist and National Audubon Society board member Paul Ehrlich has said, "Most of the ranchers in the western United States are on wel-fare, and their welfare is contributed to particularly by United States senators from the West who often sneer at welfare programs to help poor Black kids in Harlem but are happy to deliver enormous amounts of welfare larger than any Black kid in Harlem ever dreamed of, to already rich ranchers."

Financial losses are not the only costs of grazing that taxpayers must bear. Ranchers who graze livestock on public lands often become ob-stacles to sound wildlife management. For example, in the past they have blocked protective measures for threatened grizzly bears in na-tional forests surrounding Yellowstone National Park, and they con-tinue to be the biggest obstacle to the reintroduction of threatened gray wolves to Yellowstone and of endangered Mexican gray wolves to White Sands Missile Range (see *Audubon Perspectives: Fight for Sur-vival* for a more detailed account of the wolf controversy). The ranch-ers who lease public lands for grazing have come to think of those lands as theirs by right. They view wildlife conservation on public grazing lands—lands owned by all U.S. citizens and administered with federal taxes—as an intrusion on their right to grow cows and sheep. For example, in a letter to the New Mexico State Game Commission from the Catron County Cattle Growers Association, stockmen com-plained that wildlife management on public rangelands hindered the grazing and watering of cattle and expressed concern—in pseudo-le-galese—about "the taking of private property and property rights [de-fined as grazing permits on public lands] by wildlife without just com-pensation in violation of the fifth amendment to the United States Constitution." In effect, ranchers are blaming wildlife, on whose last fragments of protected habitat the ranchers graze cattle, for stealing the ranchers' inalienable rights.

This would be laughable were it not that the public grazing lands appropriated by ranchers serve as a last refuge for many wildlife spe-cies whose original habitat has been taken over by cities, suburbs, farms, and ranches. Says Tony Povilitis, a representative of the Hu-mane Society of the United States, "We have truly an endangered spe-cies crisis in the United States. We have between 1,500 and 4,500 species that are threatened or endangered. Many of those are not yet listed, they're not given protection under the Endangered Species Act. The fact is that from now and into the future, the public lands are going to be the crucial refuge for many of these species. If we don't protect wildlife there, then where?''

The grassland habitats of hundreds of species are being slowly and systematically destroyed by overgrazing. A 1985 study by the National Resources Defense Council and the National Wildlife Federation concluded that 29 percent of the Bureau of Land Management public range is in poor condition and 42 percent in fair. Only 1.9 percent was rated as excellent, and 27 percent good. Ranchers could argue that the conservation groups judge range condition harshly to make the situation look as bad as possible, but the bureau's own figures differ only slightly. They show 4 percent in excellent condition and 18 percent in poor. The other figures are nearly an exact match.

The declines of many endangered species, such as the desert tortoise, are directly linked to grazing. Cows compete with the reptiles for food and kill the young outright by stepping on them.

The Rio Grande at Espanola, New Mexico, was the site of an 1800s sheep-trail crossing. The river exhibits thick, healthy riparian (wooded streamside) habitat on both banks. Bushes and trees help control erosion, and they also attract wildlife. Sadly, much of this habitat has been lost, and continues to be lost, taking with it wildlife's chances for survival.

◀

The terrible state of public rangeland for grazing purposes means that the rangeland is also in terrible condition as wildlife habitat. Grasslands plants and animals, particularly in the Southwest, survive in an ecosystem characterized by its aridity. During most of the year rainfall is slight. When rains do come, they tend to come in deluges, concentrated within a few weeks or months of the year. Individual storms are fierce and of short duration. Moreover, relatively wet periods of several years' duration alternate with years of drought, when the land becomes parched. Western grasslands are habitats of extremes, and the creatures that live there are adapted to the quench-and-thirst cycle. The desert bighorn sheep, for example, while feeding on shrubs and dry grass and enduring peak summer temperatures of up to 115 degrees, can go up to three days without drinking. Some grasslands rodents, such as the kangaroo rat, never drink, getting their water instead as a by-product of the digestion of plant foods. Grass species adapted to these arid regions are typically deep-rooted and can withstand the withering of their above-ground portions, which regrow from the roots when rains come.

Perhaps the most ecologically sensitive portions of grasslands are the borders of streams, which are called *riparian areas*. Because rainfall is intermittent in grasslands regions, streams cannot survive without vegetative cover on the soil. Vegetation helps hold back water, letting it percolate into the earth and seep slowly toward streambeds. This slow movement of water through the soil maintains streams year-round, even during dry months.

Ranchers who graze cattle on the public lands have a long history of overgrazing. It began in the days of the open range, when each rancher—since he did not own the land and was competing with other ranchers for its maximum use—wanted to put as many cattle on the land as he could. This pattern continued into recent decades. Says New Mexico rancher Bill Cunningham, "In the old days, when I was quite young, why, we really fought numbers with numbers because we were competitive against one another, and especially in a clash of sheep against cattle. And the way you handled that was you ran more than the neighbor did. And of course, we knew what that was doing to the range."

What it did to the range was destroy the grass cover, wiping out the more nutritious grasses that provided the best cover for the land, the best means for regulating the flow of water through the soil. Wildlife dependent upon the grasses declined as the long-lived grass species were replaced by short-lived weed species of lesser nutritional value. As these were eaten, the surviving plant species tended to be those that were poisonous to livestock or that were armed with sharp, discouraging spines. Grasslands vegetation thus changed into something unnatural and unhealthy for wild creatures dependent on grasslands habitat. The land's nutritional value for livestock also was reduced, but since cattle and especially sheep are less selective about what they eat than are many wildlife species, livestock could get by. Moreover, livestock were supplemented with feed, giving them a distinct advantage over wild species.

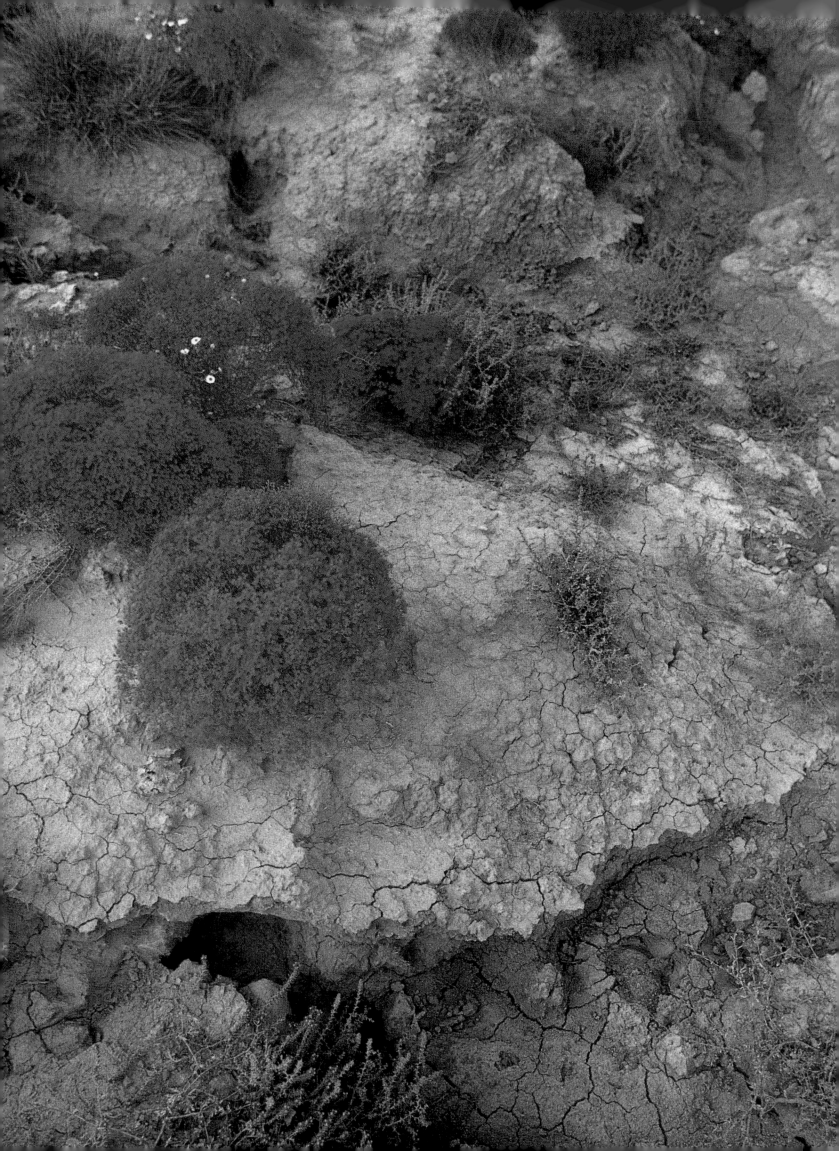

B right-yellow broom snakeweed has long roots, making it a survivor in the parched, overgrazed landscape near Rio Puerco, New Mexico. Its abundance indicates an abused landscape.

◄

Not Enough to Drink

In addition to destroying the rangeland's grassy cover, livestock grazing had another dangerous effect. Most American cattle breeds came originally from northern Europe, where they had plenty of water. Consequently, cattle, unlike native grasslands species, drink a lot. And on top of what they drink must be added the water that goes into grain crops grown for livestock feed. Of the roughly 100 billion gallons of water used daily in the United States, 84 billion are drained away by 17 western states that generally are home to more livestock than people. Much of this water comes from tax-supported water projects. Fully 90 percent of the water taken from the upper and lower portions of the Colorado River goes to crops grown for livestock. In Montana, 97 percent of water taken from rivers and streams goes to agricultural use, primarily irrigation for hay fields. The average cow in its lifetime sucks up enough water to float a Navy destroyer. Every steak you put on your plate represents the absorption into cow flesh of roughly 3,000 gallons of water.

Given the need of cattle for water, it is not surprising to find that they tend to concentrate around water sources. In many areas, ranchers and the federal government have sunk wells and put up tanks to provide water for livestock. The livestock congregate around these tanks, which are usually characterized by a perimeter of denuded muck. Beyond that lies a stubble of overgrazed land, because the animals do not like to roam far from water sources. They will feed in areas surrounding water until the grass and shrubs are gone.

Worse still, they also congregate along streams. Here the effects are even more harmful. An adult cow or bull weighs close to half a ton and with each step bears down on the ground with significant force. Herds that gather around streams quickly break down banks, making streams wider and shallower and muddying the waters. Sediment darkens the streams, causing the water to warm in the sun and threatening fish and other stream species adapted to cooler, clearer waters. As areas surrounding streams are denuded of grasses and compacted by hooves, water runs more rapidly into the streams, leading to floods after rains. Because water rushes into the streams rather than being diffused gradually through grass cover, streams often run dry. According to a 1988 Arizona Fish and Game Department report, less than 3 percent of Arizona's original riparian zones remain intact. The federal Environmental Protection Agency estimates that nearly 30 percent of the sediment reaching western streams is caused by livestock grazing, and a 1988 report by the General Accounting Office declares that "poorly managed livestock grazing is the major cause of degraded riparian habitat on federal rangelands."

Riparian damage has had deleterious effects on wildlife. Grazing may be blamed, for instance, as the chief cause of declines in the West's native trout species. According to the Arizona Fish and Game Department's 1988 report, both Arizona and New Mexico have lost 90 percent of their riparian zones, primarily to grazing. The cottonwood/willow forests that flanked the streams in pre-livestock days have been destroyed as cattle and sheep ate saplings, and old trees died without replacements. Fewer than 20 stands of cottonwood/wil-

low, once common in the Southwest, still remain in Arizona, and only 5 of those are extensive, according to the Arizona Nature Conservancy. These lost riparian habitats were important for many wildlife species. Some 75 percent of the terrestrial species in southeast Oregon's Great Basin are dependent upon, or use, riparian habitats, according to an Environmental Protection Agency report published in 1990. More than three-quarters of all wildlife species in southeastern Wyoming depend on riparian habitat. Riparian areas provide habitat for more bird species than all other western rangeland vegetation types combined, and half of all birds species in the Southwest are completely dependent on riparian habitat. The endangerment of riparian species illustrates the harmful effects of grazing on the wildlife community. In the pre-livestock era, Arizona's streams harbored 32 fish species. Five are extinct in the state, and 21 of the remaining 27 are listed as threatened or endangered or are being considered for listing. Eighty-one of Arizona's 115 threatened vertebrate species and 69 of New Mexico's 94 are closely associated with, or dependent upon, riparian habitat.

Despite the threat to wildlife, little has been done to protect these areas from livestock or to repair the damage. According to a 1988 General Accounting Office report on riparian habitat, "There are no technical barriers to improving riparian areas and . . . basic restoration used on successful projects can essentially be applied to all riparian areas on federal rangelands." However, the bureau and the Forest Service have restored relatively few riparian habitats during the past two decades for two reasons, according to the report. One, the staff needed to do the work had been severely reduced during the Reagan years. On average, each rangeland manager in the Forest Service or the bureau is responsible for roughly 191,000 acres, or about 300 square

►

miles. Consequently, monitoring of grazing allotments, riparian or otherwise, is scarcely adequate: According to another General Accounting Office 1988 report on declining rangeland, about 15 percent of all grazing allotments had not received a monitoring visit in the five years previous to the report, and about half of those not monitored "were in the intensive management category and include allotments the range managers identified as overstocked or in declining condition."

The second reason for the failure in riparian improvement was that "many field staff responsible for riparian improvement work, primarily in BLM, do not believe their work will be supported by agency management if it is opposed by ranchers using the public rangelands."

Personnel in the federal grazing agencies have good reason for their sense of futility. Often when range managers have tried to initiate reforms such as reductions in cattle numbers, they have typically found themselves transferred, at the behest of local ranchers, to new posts where their ideas will have little or no impact. According to the 1988 General Accounting Office report on grazing in riparian areas, staff in one Bureau of Land Management grazing district said they were essentially "directed by headquarters and the state office to make no decisions opposed by permittees. Further, BLM is not managing permittees; rather, permittees are managing BLM." The case cited involved an area manager who stopped a rancher he found cutting riparian trees illegally. "Soon after," the report said, "the area manager was told by his district manager that word of the incident had gotten back to him as a result of the rancher's political connections. The area manager was told to apologize to the permittee and deliver the wood to his ranch." In another case, a bureau biologist responsible for riparian programs in a field office was told by his area and district managers, when he tried to stop a permittee who was illegally putting cattle on a riparian habitat, "that they would not support his recommendations for trespass penalties or specific riparian improvements if they involved a conflict with permittee interests."

Given the public-land grazers' political clout, you might think that they are providing some national service that mitigates the massive federal subsidies and the rangeland destruction. Surprisingly, these ranchers do little or nothing for the rest of the nation's taxpayers. Though engaged in the business of producing beef, they produce too little to be of any economic consequence, beyond the tax dollars they appropriate for themselves. Critics of public-land grazing say they provide no more than 3 percent of the meat that reaches our tables. That is about 2.5 pounds of the estimated 77 pounds of beef that the average American eats each year. The grazers claim they provide as much as 15 of the 77 pounds, which still means that 80 percent of our beef comes not from the West—where arid lands are being punished to line the wallets of a few stockmen—but from the South and Midwest, the true beef states.

Not only are the public rangelands being destroyed needlessly, but, even more frustrating, the solutions to the problems are well known. Some experts even say grazing could continue without the concomitant damage if federal agencies would only adopt some very simple measures for range protection.

Solutions: Protecting Our Rangelands

Perhaps the most highly touted approach to improving rangeland management has involved fee increases. Critics would like to raise the cheap federal rate to a fair-market level. The Reagan and Bush administrations and their congressional allies, which constitute the stockmen's political power base, have generally been unmovable obstacles to congressional drives for a fee increase. But even increasing the fees, while it might provide more funds for rangeland management, still would not solve the central problem: too many cattle. Ranchers would just be paying more to continue their destructive practices.

Other solutions attempt to tackle the problem of livestock numbers. One of the most controversial of these proposals is also the one most appealing to stockmen. This is an idea pushed by Allan Savory, formerly of Zimbabwe but now a U.S. resident after fleeing his native land because of political difficulties and the failure of his game ranch during drought. Savory—who has a degree in botany—says that cattle should be used as a tool to increase grass growth. He maintains that in pre-Columbian America, bison and other hoofed animals broke up and fertilized the soil. Now that the native species are gone, cattle should do the job, and the best way to get them to do it is to double their numbers on grazing allotments or, if numbers cannot be increased, to put them into smaller pastures so they will be more concentrated. In the second case, they would be periodically rotated from

Jim Winder, Holistic Resource Management rancher near Nutt, New Mexico, drives his pickup to check on a back lot. He and his stepfather run 700 head on 18,000 acres, three-quarters of which is state and federal lands. Winder's feelings for the range are reflected by excellent grasslands and a love of wildlife, even coyotes.

A grasshopper has just shed its skin. Rancher Jim Winder says he used to worry about grasshoppers, "but then I noticed you only see them on weeds." Winder calls himself "an environmentalist's best friend," because "I'm out here doing it," protecting the range.

pasture to pasture. Doubling cattle numbers, he says, will halt desertification—the process by which overgrazed grasslands turn to desert, as is happening over much of the Southwest—and increase grass production.

Savory has found ready disciples for his Holistic Resource Management Plan, as he calls it, among the ranching community. Doubling the number of cattle is a change ranchers can support. Says Jim Winder, a New Mexico rancher experimenting with holistic ranching, "What we're trying to do is apply this herd effect to knock down old, senile grass plants and turn up the soil, cultivate the soil much like somebody'd cultivate their garden, and to incorporate the seed and other matter, organic matter, into the grass or into the soil." Some ranchers are in their fifth year of holistic grazing under Savory's direction.

Critics of the approach say that it has a number of fundamental flaws, perhaps the most fundamental being that big herds of large hoofed animals never roamed the Southwest. The bison were farther east and north, and other hoofed species, such as deer and pronghorn, were never present in large enough numbers to produce a "herd effect." Perhaps more important, experimental tests of Savory's holistic management have failed. A pilot study on the Apache-Sitgreaves National Forest in Arizona produced a test allotment in which, after four years of holistic management, there was not enough grass to get cattle through the next winter nor enough reserve for wildlife or vigorous plant growth. At a Bureau of Land Management test site in southern Arizona, five years of holistic grazing failed to show any improvement in vegetation, except that inedible snakeweed increased. Five studies in Texas and New Mexico found that the Savory approach caused a decline in the absorption of water by soil and a concommitant increase in erosion.

Jerry Holechek, a New Mexico State University professor of range sciences who has been studying grazing for some 30 years, says of Savory's theory, "We now have several studies that show that the hoof-action effect has either a negative impact or no impact at all, compared to, say, moderate continuous grazing or no grazing at all."

Holechek has his own solutions to the grazing issue. He believes that moderate levels of grazing are better for rangeland than no grazing and are certainly an improvement compared to overgrazing. In an unpublished paper he says, "Studies from the Chihuahuan desert show long-term recovery occurs as soon if not more rapidly under light to moderate grazing as under complete rest." One reason that moderate grazing helps is that it removes dead plant matter that accumulates on the ground. Writes Holechek, "Excessive accumulations of standing dead material chemically and physically inhibits new growth, and can provide habitat for insects and diseases that weaken or kill the plant."

Research shows, Holechek points out in his paper, that in heavy-use areas, where grazing removes 50 percent of the forage yearly, the production of forage "averages less than one third while biomass of broom snakeweed and other poisonous plants is over twice that on the moderately stocked . . . pastures." In areas where forage removal is kept at about 30 percent yearly, grass production is greater and "long-term cattle productivity (pounds of meat produced per cow) has equalled or exceeded that of the best ranges in north-central Texas or eastern Kansas. This is in sharp contrast [to overgrazed] Chihuahuan desert ranges in New Mexico where calf drops and weaning weights are much lower and death losses are higher." Another benefit: As the land recovers from overgrazing, the amount of forage increases, so that cattle numbers may be increased, too, as long as no more than 30 percent of the forage is taken by grazing each year.

Holechek is prograzing. "The land can be grazed without destroying the land," he says. "But at the same time, I think range people and ranchers need to understand that the amount of grazing the land can sustain is probably less than the pressure we presently have in a lot of these arid areas." To encourage ranchers to reduce the number of cattle they graze, says Holechek, they should be given an incentive, such as reduced fees in return for fewer cattle. Sierra Club field representative Rose Strickland agrees. "The problem with the current public land grazing system is that all the incentives in the system work to keep the same number of cattle on the public land. Right now there is no incentive to do a better job. The livestock grantor on this side of the fence may be doing an excellent job. The guy on the other side of the fence may be doing a lousy job. They pay the same grazing fee, they get the same grazing permit year after year. There is no incentive for the guy who's doing the good job to keep on doing a good job. There's no incentive for the guy doing a bad job to do better."

Holechek blames Bureau of Land Management policies for much of the damage. "I think it would be better without the BLM, which spends three dollars to get two dollars back," he says. "I think ranchers would have less tendency to overgraze without BLM because of that agency's inconsistency and policy changes. The range would be in as good shape if we left ranchers alone than under the present system. We have to reward good ranching with lower prices or subsidies. . . . I believe there are a lot of creative ways to reward ranchers who do a good job."

Many grazing critics disagree with Holechek's middle-of-the-road position. Some antigrazing activists in New Mexico will tell you that the ranchers have abused the land for too long, that it is time to throw them off of the public land. After viewing a stream that had been heavily eroded because of grazing, independent wildlife consultant Steve Johnson of Tucson, Arizona, said, "Cattle are an unnatural disaster in this part of the country. They never occurred here. They're an exotic, introduced ungulate with great capacity for destruction. They cannot, in my opinion, be managed in ways that are commercially profitable without this being the result in areas that are especially sensitive."

Some critics are ready to take a hard stance. One activist pointed out that many ranchers near urban areas do not live on their ranches but, rather, leave their cattle unmonitored on public lands during the week and check them only on weekends. This activist observed that a lot can happen during the owner's week of absence. A well might be shut off, which could kill a herd between a rancher's visits, or cattle might simply be shot during the week. Clearly, the controversy over public grazing is heating up.

Bessie strikes a pose at the University of New Mexico research ranch. Cows in the arid Southwest are resource gluttons, gorging themselves on barnloads of grass and barrels of water, both of which are scarce commodities. Very little of the 77 pounds of beef eaten each year by the average American comes from this overtaxed region.

▶

Jim Fish, public-lands activist, leads a tour of a sorely overgrazed Bureau of Land Management range near the Rio Puerco. On his patrols Fish finds numerous signs of neglect or outright mismanagement: severe erosion, unmended fences, noxious weeds, bullet-ridden signs, and makeshift attempts to treat the symptoms, not the causes, of overgrazing.

Jim Fish, a Princeton-educated engineer with the Department of Defense's Sandia National Laboratory near Albuquerque, New Mexico, says that cattle must be shed from the public lands. Speaking with a western accent that reveals his upbringing on a Texas ranch, Fish says, "What I see out in these rangelands are severely degraded situations where we don't have the vegetation, we don't have the wildlife that ought to be out here because of overgrazing." The solution, he says—and this is a message he reinforces in his *Grasslands Activist Newsletter,* which urges reform in public rangeland grazing—is to get rid of the cattle. Some studies support his contention that the no-cow approach is the key to grasslands recovery. One example is the Audubon Research Ranch near Elgin, Arizona. Some 20 years ago vegetation covered only 20 percent of the land. Grazing was stopped in 1969, and since then vegetation has covered 80 percent. Tall grass species have replaced short- and mid-grass species, and flowering plants have increased. The ranch has fewer grasshoppers than neighboring grazed lands because it has more birds, which find better nesting cover on the ungrazed land. Says Jane Bock, who has been studying the effect of nongrazing on the Audubon ranch, "If we look at our grassland and analyze our native grass species, there's nothing new, nothing you couldn't with a bit of hard work find on the surrounding ranches. It's just that the sort of gaps or holes that have been filled in with vegetation on our grasslands have been filled in with these native grasses and they're common here in comparison with our ranching neighbors."

Adds her husband, Carl Bock, who works with Jane as co-director of grasslands research for the National Audubon Society, "The surrounding cattle ranches here are well managed by thoughtful owners and operators, and they're really not in that bad a shape. But they've been grazed continuously for at least 125 years and you can't have all those cows eating all that grass for all those years without a significant impact. In fact, we've discovered that cows are the driving ecological force in these lands if they're present. They set the rules for what other kinds of plants and animals can occur."

Removing cattle from public lands would undoubtedly put some small ranchers out of business. They lack the land they need to sustain their operations. But Fish, the son of a rancher, sympathizes more with the dying land than with the stockmen who are killing it. Let them go out of business, if need be, to protect the land, he says. "I feel like they're a dying breed anyway. And I just don't see why we should let them continue to destroy the land, the public land, to support a lifestyle whose time has passed."

The success of any of these proposals depends ultimately on a single, underlying factor: rancher opposition. The ranchers are powerful enough politically to block any changes in grazing policies that they do not like. They have been permitted for more than a century to use public lands as if they were a personal fiefdom and have stopped virtually every measure to reduce grazing levels or to raise fees. Meanwhile, roughly half the public rangelands languish in only fair to poor condition because of the ranchers' excesses. Moreover, though the stockmen are heavily dependent on tax handouts, many of them per-

An aerial photograph taken over Cabezone, "the town that grazed itself to death," and Rio Puerco, shows appalling erosion and bank-cutting. Rancher opposition has blocked reforms that would have vastly improved the nation's public lands.

sist in resenting any outside suggestions that they properly manage the lands they use. Cloistered away in the narrow confines of their world, they have yet to understand that the regionalism of the past is as dead as travel by horseback. We are citizens not of the local area in which we live but of the entire nation. All of us have a stake in the condition of our public lands and in the use to which our tax dollars are put. The entire nation is our habitat, and we must protect it as the home that it is to us.

We face in the West an old-fashioned showdown where the weapons are not Colts and Army revolvers but science, data, and ideas. Those who want to protect and preserve our rangelands will have to provide the political clout needed to overcome the ranchers' opposition. Despite all the proposed solutions to the grazing issues, there is only one solution that will work: Speak out!

AS LONG AS THE GRASS SHALL GROW

 hundred and fifty years ago, that vast portion of the Midwest called the Great Plains was thought to be nothing more than a dry, empty place pioneers had to pass through on their way to the promised land in Oregon, Washington, and California. Today, that opinion of the Great Plains generally has not changed. The majority of the 235 million U.S. citizens who do not live on the plains—roughly 95 percent of the populace—doubtless still think of it as a monotonously flat, mind-deadening space that you have to drive through for interminable hours during transcontinental trips, or fly over as quickly as possible.

But the Great Plains is a great victim of that often dangerous pursuit we call human progress. If we could see it as it was before the French sold it to Thomas Jefferson, we would have an entirely different opinion. The plains would be a veritable tourist mecca, a global drawing card. It was a place of tremendous natural splendor, rivaled in recent history only by the savannas of Africa.

Our predecessors turned it into a food factory, fashioning what had been called the Great American Desert into the Breadbasket of the World. Probably not a single one of us in the United States, and much of the rest of the world, has not therefore been affected in some way by the Great Plains, even if we do choose to pass it by as quickly as possible.

It has always been a place of extremes. Spring often brings devastating tornados accompanied by pounding hail. Summers are hot, often topping 100 degrees. Winters are bitter, subzero with high winds and pummeling snow. Temperatures may plummet 50 degrees within a few hours. In especially bad winters, such as the record breakers of the mid-1880s and in the late 1940s, the thermometer can drop as low as 50 degrees below zero. In summer, a morning of flawlessly clear skies may be followed by a torrential cloudburst that flattens crops in a pounding deluge, to be succeeded in a few minutes by more sunshine. As they say throughout the region, if you don't like the weather, wait a few minutes; it will change.

Rainfall is sporadic throughout the warmer months of the year. It comes in floods during spring and disappears for weeks, sometimes months, during the hottest seasons. Annual rainfall amounts to about 12 to 22 inches. Native grasses are adapted to a short period of growth after spring rains, followed by a drought rest period in summer.

The plains have always been a tough place to live in. The first European residents found that they had to build their houses from dirt and heat them by burning dried bison dung, not the sort of conditions to recommend the place to the most discerning pioneer. Nevertheless, they came, they saw, they plowed, making the prairie perhaps our most abused ecosystem. They also turned it into one of the most powerful agricultural machines in the world. The mottos and symbols that have appeared over the years on midwestern license plates give some idea of the pride the residents take in this achievement: Nebraska's

◀

Fall adds patches of color to the Flint Hills of Kansas. In this part of the country, tallgrass prairie predominates, and trees and other woody vegetation are limited by fire and lack of rain to streamsides and riverbanks.

license plates have borne two mottos in recent memory—"The Beef State" and "The Cornhusker State." Kansas plates once featured a graphic of golden stalks of wheat. When you enter Nebraska by interstate highway, the big sign heralding the event informs you that you have come to the home of "the good life."

For many who farm and ranch in the Great Plains, it *is* a good life. Hard, perhaps, and hardscrabble—right on the economic edge—but nevertheless clean and good, the heart of the heart of the country, where the old traditions of God and patriotism are strong and little questioned, where the skills needed for hard physical work are still a source of pride, where the day begins with the sun and ends at dark, where the air is clean and the roads empty.

But is the bottom falling out of the breadbasket? It is woven of two seemingly frail reeds—high-tech farming and irrigation, both of which may fail within the next few decades.

GREAT PLAINS CROPLANDS

THE ISSUE The Great Plains region is part of the greatest crop-production machine in the world. But systematic ecological abuses are jeopardizing not only human health, wildlife, and the natural environment, but also the agricultural industry itself.

THE CAUSE Agriculture in the Great Plains is heavily dependent on irrigation. Much of the water for irrigation comes from the Ogallala Aquifer, a vast underground lake. But water is being drawn out so rapidly that irrigation cannot be sustained in large parts of the plains at present rates for more than two or three more decades. Use of fertilizers and pesticides is contaminating waterways and other sources of drinking water. Erosion is wasting away the land and contaminating waterways. Airborne toxins are invading distant areas, such as the Great Lakes. Wetlands are being drained for crops.

EFFECTS ON WILDLIFE Wildlife suffers under the impact of pesticides and fertilizers through contamination of aquatic systems and loss of food sources and shelter.

EFFECTS ON HUMANS Contamination of drinking water may cause cancers, birth defects, and childhood fatalities. Costs of conventional farming lead to burdensome taxpayer subsidies. Crops treated with fertilizers and pesticides often carry toxic residues that pose threats to health.

EFFECTS ON HABITAT Loss of wetlands reduces breeding and living areas for a large range of species, as does the clearing of grasslands for crops.

From desert to breadbasket to basket case—will this be the story of the Great Plains? And if so, can we live without it?

A Bit of History

The first settlers of the Great Plains were nomadic hunters whose ancestors had reached North America from Asia by hiking the land bridge that connected Alaska to Siberia during the most recent Ice Age. By the time the bridge vanished under the sea some 15,000 years ago, these peoples were hunting mammoths, giant sloths, and a giant bison species as well as other large mammals, all now extinct. A portion of these people became the first settlers of the Great Plains.

Of course, they left no written records, such as land claims. Nor did they develop imposing armies. Consequently, their descendants could muster little resistance when Spanish troops traveling under the command of Francisco Vásquez de Coronado, a provincial governor from Mexico, reached Kansas in the sixteenth century while looking for the Seven Cities of Gold. He found no gold, as we all well know, but he did discover a vast, treeless plain where rivers were few and far between and where there were no good landmarks for orienting explorers in new areas. Coronado's troops were frequently lost, sometimes for days, sometimes forever. They found it very hard to backtrack because the resilient grasses showed little sign of their paths. "Although the grass was short," wrote one of the men, "when it was trampled it stood up again as clean and straight as before." The only trees, he reported, grew along the rivers.

They found plenty of bison, which they often referred to as cattle. "Traveling over the plains, there was not a single day until my return that I lost sight of them," Coronado later told the king of Spain. The Spaniards left us the first description of a bison stampede. It occurred after they frightened a herd into a blind panic that sent the animals crashing into a deep ravine.

> So many cattle fell into it that it was filled and the other cattle crossed over them. The men on horseback who followed them fell on top of the cattle, not knowing what had happened. Three of the horses that fell disappeared, with their saddles and bridles, among the cattle, and were never recovered.

Mandan chief Mató-Tópe, or Four Bears, was painted in 1833 by Swiss artist Karl Bodmer. The skin markings and feathers represent acts of battlefield bravery or war wounds.

Formal exploration of the region began with a French fur trader named Pierre Gaultier de Varennes, Sieur de La Vérendrye, who traveled into the northern plains around 1740. He encountered Mandan Indian villages scattered for miles along the Missouri in parts of North and South Dakota, some of them so big that his men frequently became lost in them. But La Vérendrye left only a poor account of his travels, written after his return home following an expedition of some five years duration. Even the names he gave to the Native American peoples he met along the way are indecipherable today. His main purpose was to pass through the plains and find the Pacific Ocean.

Success in crossing the prairie to the Pacific remained to be enjoyed by Meriwether Lewis and William Clark, whose expedition beginning in 1803 was mounted during Thomas Jefferson's administration to explore the region encompassed by the Louisiana Purchase. They left us some of the earliest descriptions of plains wildlife and the Native Americans living on the plains, but they were not the first white explorers to leave their mark upon the region. Various traders and trappers were there to meet them. Smallpox, brought in by the whites, had gone before them, too, so they found many of La Vérendrye's Mandan villages in ruins, the people nearly wiped out.

Soon after Lewis and Clark, what must have been among the first ecotourists started dribbling into the West. Artist George Catlin, a native of Pennsylvania, abandoned life as an attorney to go west and travel among the Native Americans and their wild country. In the 1830s he sailed by steamship up the Missouri from St. Louis. He loved the rolling grasslands that eased up to the river from all sides:

> One thousand miles or more of the upper part of the river, was, to my eye, like fairy-land; and during our transit through that part of our voyage, I was most of the time rivetted to the deck of the boat, indulging my eyes in the boundless and tireless pleasure of roaming over the thousand hills, and bluffs, and dales, and ravines; where the astonished herds of buffaloes, of elks, and antelopes, and sneaking wolves, and mountain-goats, were to be seen bounding up and down and over the green fields. . . .

In *The Buffalo Hunt: Surrounding the Herd*, a Currier and Ives print of a George Catlin painting, the bravest hunters ride into the teeming bison to spur them into panic. Buffalo once roamed the plains by the millions, but market hunters had depleted the herds by the late 1800s.

At the time that Catlin first reached the northern plains, writer Washington Irving was heading into the southern plains. In 1832 the 49-year-old author traveled from the East by boat and wagon to Fort Gibson in what is now Oklahoma. In his book *A Tour of the Prairies,* he left us a superb account of life on the untrammeled plains. It was a world we will never know. After chasing bison in what is today central Oklahoma, a gateway to the great plains, Irving contemplated that world:

> To one unaccustomed to it, there is something inexpressibly lonely in the solitude of the prairie. The loneliness of a forest seems nothing to it. There the view is shut in by trees, and the imagination is left free to picture some livelier scene beyond. But here we have an immense extent of landscape without a sign of human existence. We have the consciousness of being far, far beyond the bounds of human habitation; we feel as if moving in the midst of a desert world.

But even at that early date, when settlement of the plains had yet to begin, the first signs of a new era were evident. Irving wrote that he "always felt disposed" to linger at each campsite when his party moved on in the morning. He would wait, a solitary figure, in the abandoned camp "until the last straggler disappeared among the trees, and the distant note of the bugle died upon the ear, that I might behold the wilderness relapsing into silence and solitude." The campsites, established when possible in stands of trees or in woods along rivers,were forlorn and desolate.

> The surrounding forest had been in many places trampled into quagmire. Trees felled and partly hewn in pieces, and scattered in huge fragments; tent-poles stripped of their covering; smouldering fires, with great morsels of roasted venison and buffalo meat, standing in wooden spits before them, hacked and slashed by the knives of hungry hunters; while around were strewed the hides, the horns, the antlers and bones of buffaloes and deer, with uncooked joints, and unplucked turkeys, left behind with that reckless improvidence and wastefulness which young hunters are apt to indulge when in a neighborhood where game abounds.

The entire nation was, in a sense, made up of such young hunters in those days, people searching for opportunity and riches. One of the more memorable opportunities arrived in 1849 with the discovery of gold in northern California. The movement west—a mere trickle in the 1840s, became a deluge in the 1850s as wagon trains pushed across the prairies for the West Coast. These people were just passing through. Some found the dry, treeless plain too much for them and turned back. Others warmed to it. Wrote wagoneer Lydia Allen Rudd in her diary entry for May 11, 1852,

> We met several [travelers] that had taken the back track for the states homesick I presume let them go We have passed through a handsome country and have encamped on the Nimehaw river the most beautiful spot that ever I saw in my life I would like to live here As far as the eye can reach either way lay handsome rolling prairies not a stone a tree nor a bush even nothing but grass and flower meets the eye until you reach the

valley of the river which is as level as the house floor and about half a mile wide, where on the bank of the stream for two or three rods wide is one of the heaviest belts of timber I ever saw covered with thick foliage so thick that you could not get a glimpse of the stream through it.

The Native Americans of the plains were feared by the wagoneers but, in the 1850s, generally were not hostile. They were mainly the victims of bad publicity. An 1869 federal commission on Indians noted,

The history of the Government connections with the Indians is a shameful record of broken treaties and unfulfilled promises. The history of the border, white man's connection with the Indians is a sickening record of murder, outrage, robbery, and wrongs committed by the former, as the rule, and occasional savage outbreaks and unspeakably barbarous deeds of retaliation by the latter, as the exception. . . . The testimony of some of the highest military officers of the United States is on record to the effect that, in our Indian wars, almost without exception, the first aggressions have been made by the white man, and the assertion is supported by every civilian of reputation who has studied the subject.

The Native Americans encountered by early wagon trains were primarily interested in trade. The farmers who were moving west were often poor hunters and depended upon the Native Americans for food. Indian attacks were rare. Exposure to the elements, lack of food, and disease were greater dangers. "We had a dreadful storm of rain and hail last night and very sharp lightning," wrote Amelia Stewart Knight on May 17, 1853.

It killed two oxen for one man. We have just encamped on a large flat prairie, when the storm commenced in all its fury and in two minutes

Indians hunt on horses covered with bison skins, in a painting by Frederic Remington, published around 1908. The lands of Native Americans were appropriated by the government and their food supply was eliminated by market hunters, leaving the once self-sufficient tribes dependant on assistance programs.

after the cattle were taken from the wagons every brute was gone out of sight, cows, calves, horses, all gone before the storm like so many wild beasts. I never saw such a storm. The wind was so high I thought it would tear the wagons to pieces.

Cholera killed many on the plains. The wagon trail that paralleled the Platte River in Nebraska was studded with grave markers. Lydia Allen Rudd left the following record of her 1852 crossing:

> May 14 Just after we started this morning we passed four men digging a grave They were packers The man that had died was taken sick yesterday noon and died last night They called it cholera morbus. . . . We passed three more graves this afternoon.
> May 18 Passed by another newly made grave.
> June 1 a good many sick and a number of deaths. . . .
> June 3 passed four graves this morning that the folks died yesterday all out of one train and others sick.
> June 12 passed five graves this morning and a camp where one of their men was dying.
> June 14 passed two graves today death caused from the cholera
> June 17 one young lady died last night and the other cannot live but a few hours longer both sisters Their father an old feeble gray headed man told me that within two weeks he had buried his wife one brother one sister two sons in law and his daughter died last night Past and other that cannot live the day out. . . .

Bison, symbol of the West, still roam the range on one of its last outposts, the National Bison Range in Montana. But they roam beyond bounds at the risk of being killed. Many ranchers dislike bison, which they believe transmit the disease brucellosis, a bacterial sickness first given them by an exotic, four-footed ungulate, the European bovine.

June 21 the young Mr. Girtman died today about four o clock his brother not as well Baker worse of the two I do not know when we shall be able to travel any more It begins to look rather discouraging.

It begins to look discouraging. Is it any wonder that one woman wrote during her crossing of the plains, "the heart has a thousand misgivings & the mind is tortured with anxiety & often as I passed the fresh made graves, I have glanced at the side boards of the waggon, not knowing how soon it might serve as a coffin for some one of us"?

The era of the wagon trains came to an end after the Civil War, when the first railroad line crossed the plains. Other lines followed, and the railroads soon offered special rates to settlers. The cheap fares and the accessibility to remote places that the railroads provided brought more and more settlers into the West. Various federal programs, such as the Homestead Act of 1862, helped too, offering cheap land to those willing to build homes and farms.

Old World Meets New

By the 1870s, the vast plains were showing the effects of settlement. The bison were vanishing, as were huge herds of pronghorn antelope, elk, wild sheep, and deer. The wolf and grizzly were being squeezed out. And the Native Americans were taking a pounding.

In the earliest days of the westward movement, the Native Americans of the plains were frequently willing to help the wagon trains. They traded for food, helped the settlers across rivers, and sometimes served as guides. In one of the great miscalculations of history, the Native Americans thought the settlers were just passing through. By the time they realized otherwise, it was too late to do anything about it.

Of course, technology was against the Native Americans from the outset, so their downfall was inevitable. The settlers had both the guns and the numbers. The repeating rifle, like the railroad, speeded the settlement of the plains. With rifles the military and the market hunters could rapidly wipe out the bison, which was the primary food source of the Plains Indians as well as the basis for much of their social and economic life. With the bison gone, the plains peoples were quickly reduced to vassals of the American state. Dependent upon the government for food and other goods, they became increasingly powerless. Meanwhile, corrupt officials grew wealthy selling government goods meant for Native Americans. Reservations often received less than half the food and other supplies they were promised.

The Plains natives lost their culture and their livelihood. Their efforts to fight back were futile. They could find no redress for their complaints about lack of food and unkept promises. They were a broken people, shunted about from site to site as one treaty after another was violated. They were promised Oklahoma, but it was taken from them. They were promised lands in the Dakotas—the sacred Black Hills—but those too were taken away, opened up to gold prospectors. One tribal representative, his name lost to history, told an 1876 Indian-affairs commission from Washington,

I am glad to see you, you are our friends, but I hear that you have come to move us. Tell your people that since the Great Father [the president of the United States] promised that we should never be removed we have been moved five times. I think you had better put the Indians on wheels and you can run them about wherever you wish.

By then, Native American resistance was almost over. The loss of will is perhaps symbolized in two contrasting remarks made by the great Sioux leader Sitting Bull, one of the masterminds of Indian success at the Battle of Little Bighorn in southeastern Montana, where military hero and possible presidential contender George Armstrong Custer met his fate. In the winter of 1876–77, when his people were freezing and hungry at the refuge they had sought in Canada after Little Bighorn, Sitting Bull told a friendly trader that he nevertheless refused to surrender

because I am a red man. If the Great Spirit had desired me to be a white man he would have made me so in the first place. He put in your heart certain wishes and plans, in my heart he put other and different desires. Each man is good in His sight. It is not necessary for eagles to be crows. Now we are poor but we are free. No white man controls our footsteps. If we must die we die defending our rights.

But by 1881 Sitting Bull had lost his fighting edge and told the commander of Fort Buford, to whom he surrendered,

I do not come in anger toward the white soldiers. I am very sad. My daughter went this road. I am seeking her. I will fight no more. I do not love war. I never was the aggressor. I fought only to defend my women and children. Now all my people want to return to their native land. Therefore I submit.

The Plains Indians, like Sitting Bull, were also in ruin by 1881. As Sitting Bull himself observed, life in the hands of an alien government led to deterioration. "It is bad for young men to be fed by an agent," he said. "It makes them lazy and drunken. All agency Indians I have seen were worthless. They are neither red warriors nor white farmers. They are neither wolf nor dog." But then a strange new force arose among them that would lead to the death throes of the Native American culture.

In January 1889 a Paiute named Wovoka, living on the Walker River Reservation in western Nevada, had a fever and fell into delirium just as the sun went into eclipse. When he recovered, he said he had had a vision in which he spoke with God and saw a place where dead Indians were living in peace, with many buffalo. He told any of his people who would listen that they must live in peace, for soon the old ways would be restored. "Do not tell the white people about this," he said.

Jesus is now upon the earth. He appears like a cloud. The dead are all alive again. I do not know when they will be here; maybe this fall or in the spring. When the time comes there will be no more sickness and everyone will be young again. Do not refuse to work for the whites and do not make any trouble with them until you leave them.

November hoarfrost forms on ragweed at the Lillian Annette Rowe Sanctuary on the Platte River, near Hastings, Nebraska. Until most grasslands were converted to croplands, they were among the largest terrestrial ecosystems, second only to forests. Prairies make up nearly a quarter of the land area of the Lower 48 states but are home to only about 2 percent of the population.

◄

Wovoka also taught the ghost dance to his followers, who came from as far away as the Great Plains as Wovoka's message spread through the native peoples' grapevine. Those who performed this dance, it was believed, would enter the new world. Those who did not—the whites for instance—would vanish with the old world.

Desperate for hope, Native Americans embraced the new movement. They made ghost shirts that they thought made them impervious to bullets. They danced the ghost dance. And they inadvertently alarmed the military, who suspected that some kind of uprising was on the way, especially when, in late 1890, large numbers of Sioux dancers retreated to the badlands in the northwestern reaches of the Pine Ridge Reservation. The dance had its militant aspects, and the Sioux had, after all, been a vital factor in Custer's destruction only a few years before. Now the word was being passed around—and it must have seemed ominous to the cavalry—that a new world order, devoid of the Wasichu (white men), was coming in spring 1891.

Despite the promise of Wovoka, spirits were low, even among militants. Sitting Bull was killed in December 1890 during a bungled arrest, and the Sioux were in poor shape for pitched battles. They had weathered a drought the previous summer, they were destitute, they were demoralized. By the end of December, various bands of ghost dancers were trickling back to the reservation.

The Last Resistance

The ghost-dance movement led to the final battle—if a one-sided slaughter can be called a battle—of the Indian wars. One of the last Sioux bands to return to Pine Ridge was led by Big Foot, also known as Spotted Elk. He was deathly sick with pneumonia, could not even travel by foot, and had to be carried in a wagon. His band of 350 included perhaps 100 men. On December 28, 1890, the band was intercepted by eight troopers from the Seventh Cavalry—Custer's old command—about 30 miles from the Pine Ridge Agency. The band was cold and hungry. It offered no resistance when the troopers said they would escort the Sioux to the army camp at Wounded Knee Creek. Big Foot was given a tent and stove.

When the Indians awoke the next morning, they found themselves surrounded by 500 troopers, including four Hotchkiss cannons, also called wagon guns, which could fire 50 explosive shells per minute. The soldiers started to take all guns away from the Sioux. This alarmed the Indians, who feared that the cavalry intended to shoot them once they were defenseless. A medicine man called out that all warriors who wore ghost shirts should be brave, because bullets would fall off of them. When troopers came to Big Foot's tent, they attempted to take rifles from two men standing by the tent door and covered from head to foot in white sheets, like Halloween ghosts. One of these Sioux, Yellow Bird, refused to turn over his weapon, and a struggle ensued, during which the gun went off and the trooper was killed. The Indians raised their arms to heaven, a devotional gesture, and then went for their weapons. The lieutenant in charge gave his men the order to fire.

Of course, most of the Sioux had been disarmed, so they had to race to where their rifles were piled, fighting bare-handed as they went, outnumbered five to one. A slaughter followed from which women and children were not excluded. Big Foot was killed at the start of the battle by a soldier who entered his tent, found him sick in bed, and shot him in the chest. Yellow Bird sought refuge in a teepee from which he fought with his rifle until the teepee caught fire. He was then shot and killed.

The death toll for the battle is unknown. Twenty-five soldiers died, some as a result of their own crossfire. A military party buried 146 Indians, more were taken from the field by relatives, and some Sioux escaped.

Charles Eastman, the physician at Pine Ridge and himself a Santee Sioux, traveled with a photographer from Chadron, Nebraska, and the burial crew. They found the first body three miles from the battle site. "From this point on, we found them scattered along as they had been relentlessly hunted down and slaughtered while fleeing for their lives." Snow had fallen after the fight, blanketing the battleground and turning bodies into white mounds. Several survivors were pulled from the snow, most of them infants who had been shielded by their mothers' dead bodies. Eastman, who recalled struggling to keep his composure, said it was "a severe ordeal for one who had so lately put all his faith in the Christian love and lofty ideals of the white man." The photographer captured the carnage on film, producing the only known photos of an Indian battleground, frozen bodies and limbs protruding from the snow.

As recounted in *Black Elk Speaks*, the autobiography of a nine-teenth-century Ogallala Sioux medicine man, some warriors searched for, and fired upon, some troops the next day. But it was a desultory last strike, for they knew that further resistance was futile. Red Cloud, an ancient warrior/chief, encouraged them to give up, and the Sioux finally surrendered. Wrote Black Elk years later,

> And so it was all over. I did not know then how much was ended. When I look back now from this high hill of my old age, I can still see the butchered women and children lying heaped and scattered all along the crooked gulch as plain as when I saw them with eyes still young. And I can see that something else died there in the bloody mud, and was buried in the blizzard. A people's dream died there. It was a beautiful dream.

New dreamers were coming, though. Settlers were already pouring into the Great Plains. No one remained there now to resist them, and some technological advances helped them along. Historically, the plains were too dry for crops and the grasslands too tightly rooted to cut through with traditional plows. But by the turn of the century, modern irrigation systems capable of drawing water from underground solved the water problem, and equipment powered by engines rather than animal muscle, particularly tractors that pulled plows, brought the wild plains country to its knees. New riches were wrested from the fertile soils of the plains. The early explorers' Great American Desert became the songwriter's amber waves of grain and fabled fruited plain. The dream turned to reality, and we enjoy its bounty every day.

But the bounty is founded on government economic policies that lead to unsustainable agricultural practices and to profound ecological abuses, abuses long ignored. It is not likely, however, that we can ignore them much longer if we wish to sustain our seemingly boundless food supply.

The Collapsing Cornucopia

U.S. agriculture produces phenomenal quantities of foodstuffs. In 1990 our farmers harvested nearly 8 billion bushels of corn, 2.7 billion bushels of wheat, 357 million bushels of oats, and 419 million bushels of barley. Of those amounts, the 10 Great Plains states—North and South Dakota, Nebraska, Montana, Wyoming, Kansas, Colorado, New Mexico, Oklahoma, and Texas—accounted for 1.8 billion bushels of corn, 1.7 billion bushels of wheat, 122 million bushels of oats, and 235 million bushels of barley (since these figures are for the entire states, not just for the Great Plains portions of each, they do not represent Great Plains production exclusively). The Department of Agriculture, from which all these figures originate, reported in 1989, the most recent year for which figures are available, that the nation's wheat harvest was worth $7.7 billion, and corn for grain (as opposed to corn grown for animal food) $17 billion. To these amounts the Great Plains states contributed $4.2 billion in wheat and nearly $8 billion in corn.

Yet, despite the big-bucks factor, the crops are costing taxpayers large sums of money in farm subsidies and other federal handouts.

Harvesters bring in the sheaves near Holdrege, Nebraska. Growing pressures to produce larger harvests for export have led farmers to till more land. Today more than six billion tons of soil are lost yearly to wind and water erosion, a figure probably exceeding that of the dustbowl years.

During the 1980s, federal farm subsidies, crop insurance, and credit cost more than $189 billion. These infusions of federal cash are, to a large degree, the keystone of the farmland economic superstructure. In Iowa—a state outside the arid Great Plains region where farmland is of a higher quality—direct subsidies account for more than 80 percent of net farm income.

On top of the federal-aid costs must be added a legion of environmental degradations that sap the nation of more billions of dollars. Perhaps the most damaging of the ecological threats is water pollution. The Environmental Protection Agency, in its 1986 *National Water Quality Inventory,* identified 47 states in which agricultural runoff was a principal reason that waterways were unable to sustain such uses as swimming and fishing. Agriculture is the nation's largest source of nonpoint-source water pollution, which accounts for half of all our water pollution problems (nonpoint sources are those such as runoff from agricultural fields or urban streets; point sources are individual sites, such as a factory or sewage-treatment plant).

Agricultural runoff pollution sends an influx of soil, pesticides, fertilizers, and other contaminants into rivers, streams, and lakes. Our agricultural industry uses 346 billion pounds of fertilizers yearly and 430 million pounds of pesticides. Many farmers are locked into using costly and dangerous chemicals because, when they are seeking advice on how to improve crop production, the most readily available sources of information are salesmen of the agrochemical industry. Federal consulting programs, too, frequently work hand in hand with the agrochemical industry.

Fertilizers add nutrients to water bodies, stimulating the growth of algae and other organisms that use up available oxygen, choking the life out of lakes and streams (see the chapters on the pollution of beaches and the Great Lakes). Pesticides threaten the integrity of natural waters, destroying aquatic organisms and making some waters unfit for human consumption. In 1989, the U.S. Geological Survey detected pesticides in 90 percent of the streams tested in 10 midwestern states. Agricultural pesticides even turn up in treated water delivered to homes by municipal waterworks as a study in one agricultural state adjcent to the Great Plains shows. Iowa's Department of Natural Resources recently tested a small sample of state drinking water supplies and found pesticides in more than 60 percent of surface-water samples that *had been treated* for such contaminants.

Surface waters are not the only problem. Groundwater—water contained in vast natural chambers beneath the Earth's surface, from which it is drawn by well pumps—is tapped for drinking water in many states. It is the primary source of drinking water for some 105 million people across the nation. Ninety-seven percent of rural residents drink from wells. In 1988, an EPA examination of wells in 26 states detected 46 pesticides that originated from farms. Atrazine, a weed killer classified by EPA as a possible carcinogen and the second most widely used herbicide in the United States, was found in the groundwater of 13 states. Alachlor, a weed killer and probable carcinogen banned in Canada after it was discovered in drinking water in southern Ontario, was found in the groundwater of 12 states. Sixteen

Ammonia, an excellent source of nitrogen, is used extensively in America's breadbasket. Unfortunately, some farmers apply their chemicals under less-than-ideal conditions, resulting in little benefit to the soil and much destruction to the environment. Groundwater pollution from agricultural chemicals is a growing problem in the Midwest.

states have aldicarb, the most acutely toxic pesticide registered by EPA, in their groundwater. Is it any wonder that farmers, according to a study by the National Cancer Institute, show higher-than-average rates of malignancies for Hodgkin's disease, leukemia, non-Hodgkin's lymphoma, multiple myeloma, and cancers of the lip, stomach, prostate, brain, and connective tissue?

Ironically, some weed killers used to keep unwanted plant growth out of farm fields actually limit a farmer's choice of crops. Atrazine is used on half of all corn acreage grown in the north central states. It is one of the most-used herbicides in Nebraska, where farmers in the central part of the state spray weed killer on 94 percent of the acreage they put into corn. But persistent atrazine residues damage any small-grain crops, soybeans, and alfalfa planted in treated fields the following year. Consequently, atrazine locks farmers into growing corn year after year in the same fields. This exhausts soil nutrients, creating a dependence on chemical fertilizers, and makes the fields more susceptible not only to insect damage but also to the invasions of wild plants (which farmers think of as weeds, the object of the spray in the first place).

One telling story was related by noted agricultural expert Frederick Kirschenmann in an article for the Winter 1988 issue of the *American Journal of Alternative Agriculture*. A cattle rancher in Montana noticed that leafy spurge, a plant cattle refuse to eat, was invading his rangeland. On the advice of some chemical experts, the rancher hired a helicopter to spray his fields with a weed killer. This cost him $25,000, but it caused the spurge to curl up. The following spring it was back again, though, and he sprayed a second time for another $25,000. The next spring the spurge was back, though in lesser numbers, but his pine trees were dying and his cattle were coming down with diseases he had never seen before.

Fertilizers are another source of contamination. Nearly 25 percent of drinking-water wells tested in Kansas and South Dakota exceeded the federal health standard for levels of nitrates, which are derived from fertilizers. Prolonged exposure to nitrates can lead to infant deaths from methemoglobinemia (oxygen loss in the blood, or blue-baby syndrome). Contaminated well water killed two infants in South Dakota in 1986.

Soil erosion itself is a problem. About a fifth of U.S. cropland is subject to serious erosion, dumping nearly a billion tons of soil each year into the nation's waterways. Silt in runoff clouds lakes and streams, making them unlivable for aquatic organisms that demand clear water. It also results in dredging costs for waterways as well as losses in recreational activities and damage to flood-control projects and to water-treatment plants. The Conservation Foundation estimated the costs of sediment damage from cropland erosion at $2.2 billion yearly. But the figure is hard to pin down because so many factors are involved, including not only damage from soil erosion itself, but also loss of soil nutrients and other elements. The U.S. Department of Agriculture puts the estimate as high as $8 billion, and other estimates have gone as high as $18 billion.

The best-known instance of soil erosion in the United States occurred on April 14, 1935. It was heralded in the Great Plains regions of the Oklahoma Panhandle by an immense black cloud that boiled out of the north on a rush of wind, stretching miles high and all across the horizon, blotting out the sun and plunging a clear spring day into the darkness of night.

It was a cloud of dust, and it was the harbinger of the Dust Bowl. The cloud had been fed and nourished by the many sodbusters of Colorado, New Mexico, Nebraska, Kansas, Oklahoma, and Texas who had come into the high plains to turn grasslands into wheat fields. In the 1920s, wheat was selling at a premium and land in the plains region was cheap—as little as $15 an acre when wheat was selling for a dollar a bushel and you could grow 20 bushels an acre. You could take a mortgage on a farm and pay it off in a single year. Grasslands were ravaged by plow throughout the region. In the Texas Panhandle alone, acreage in wheat went from 82,000 in 1909 to 2 million in 1929. Farmers tore up grasslands with plows pulled by tractors, the new miracle machines. And farmers expected miracles to happen. The plows in use, called one-way plows, turned the crust of the soil into a fine powder. This was part of the miraculous scheme: Plowing, it was widely believed, would have some sort of climatological effect that would result in regular and increased rainfall—ending the cycle of alternating drought and rain that typifies the Great Plains. The powdered earth that marked the wake of the plow would act like an insulator, keeping moisture in the soil, around the roots of crops.

It seemed to work. Rain was abundant in the 1920s and on into the 1930s. In 1931 rain fell in profusion all over the plains. The harvest

Townspeople watch as a May 1937 dust storm approaches Elkhart, Kansas, in this photograph taken by a Farm Security Administration photographer. Winds carried midwestern dust all the way to Washington, D.C.

jumped 150 percent. Then ecological factors and economics struck with a vengeance. The rain quit, as anyone who had not fallen victim to wishful thinking would have expected, and prices fell from a dollar a bushel to 25 cents. Farmers went broke and abandoned their farms. The powdery soil they had created rose into the air like a dark, shapeless specter of death, and the Dust Bowl was given birth by the plow, fathered by avarice.

Vanishing Essentials

The story of the Dust Bowl, of the desperation and poverty it spawned, is well known. John Steinbeck turned it into a political tract disguised as a novel, *The Grapes of Wrath*. Less known, perhaps, is that many of the conditions that led to the Dust Bowl are with us now. November through May is wind-erosion season in the Great Plains, and the erosion that results is monitored by the federal Soil Conservation Service. During the 1990–91 season, the service reported, 8,181,120 acres were damaged by wind, nearly half a million more than the previous season and 2.1 million more than the average for the previous 35 years. Of that 8-million-plus acres, 91 percent was cropland. Nationwide, wind erosion blows away roughly 2 billion tons of soil yearly.

According to the service's "Final Wind Erosion Report—Great Plains 1990–91 Season," "Continuing drought which resulted in insufficient cover (vegetation and snow), and high winds were reported in explanation of the damage. One windstorm on March 12 in Kansas damaged about 1.7 million acres." Moreover, the service reported,

A farmer and his son in Cimarron County, Oklahoma, raise their fence to keep it from being buried. The photograph was taken in April 1936 by Arthur Rothstein for the USDA's Farm Security Administration documentary project.

"Throughout the Great Plains, 16,914,217 acres were reported in condition to blow. This is 22% higher than the 35 year average. . . . The Northern [plains] states reported a total of 9,311,430 acres in condition to blow. . . . The Southern [plains] states reported a total of 16,609,645 acres in a condition to blow."

This condition was created in large part by federal legislation that, until 1986, provided subsidies and tax benefits to farmers who turned grasslands into croplands. In the 1970s and 1980s, capital-gains tax benefits speeded the destruction of many different land types. As noted in *Alternative Agriculture*, published by the National Research Council, "Favorite capital gains treatment provided incentives to purchase highly erodible fields and wetlands, rangelands, or forestlands at relatively low prices; convert these lands to cropland; sell them at a profit; and exclude 60 percent of the gain from taxation." Fortunately, tax-law revisions in 1986 eliminated tax incentives, and the 1985 Farm Act prohibited subsidies to farmers who plow previously unused land (sodbusters) or who drain wetlands (swampbusters).

Investment tax credits also promoted the use of groundwater for irrigation throughout the Great Plains. Much of the groundwater being tapped lies in the Ogallala Aquifer, a vast underground lake that underlies 200,000 acres of the middle states, from South Dakota to Texas. Under Nebraska alone is enough water—2 billion acre-feet—to flood an area the size of the lower 48 states a foot deep. About 2 million people drink from the aquifer, and about 170,000 wells pump up enough water to irrigate 15 million acres, more than 20 percent of all U.S. irrigated land.

We have been tapping the Ogallala for nearly a century. Initially, withdrawal was by windmills from relatively shallow wells dug by hand. Greater depths were reached after 1900 when drilling methods improved and the gasoline engine came into use to power pumps, opening more regions to withdrawals. The subsequent development of turbine engines permitted even deeper wells. Then, in the 1960s, the center-pivot irrigation system came into vogue.

The center-pivot resembles a giant lawn sprinkler. It taps water from the aquifer and runs it through a pipe or hose that is at least an eighth mile long with sprinklers attached at intervals along its length. The sprinkler pipe around a central the anchor, watering crops the way an urbanite's lawn sprinkler waters a yard, except that the center pivot can cover 133 acres or more with one revolution. The center-pivot system vastly increased the popularity of irrigation in the Great Plains. Irrigating through canals that run via gravity into fields requires a great deal of labor and does not work well in the plains because the porous soil that characterizes much of the region makes it difficult to distribute water from canals evenly all the way across a field. The center pivot solved these problems and gave rise to the round fields you see when you fly over the Great Plains. The center pivot cannot reach into the corners of quadrangular fields unless specially equipped, so corners are left to wild plants or to crops that can be dry-farmed.

The center pivot operates efficiently on slopes and sandy soils and consequently has been widely used in the Nebraska Sandhills. Coarse-grained, sandy soils hold less than an inch of water per foot of soil depth, at best only half of what fine-grained, loamy soils can hold. The center pivot applies water lightly and keeps it coming, keeping enough moisture around roots to allow crop growth in sandy soils. A 1976 study at the University of Nebraska showed that Sandhills pastures irrigated with center pivots produced 700 to 900 pounds of live beef per acre per year, compared to 27 pounds for open ranges.

About 24 million acre-feet of water are drawn out of the Ogallala Aquifer each year. Water from precipitation and other sources seeps through the soil and back into the aquifer at the rate of about 3 million acre-feet yearly. Thus, water is being withdrawn at about eight times the rate that the aquifer is recharged. Since irrigation is the main pillar of the Great Plains agricultural edifice, the draining of the aquifer threatens agriculture as we know it in a region that supplies roughly 15 percent of our corn and wheat, 25 percent of our cotton, and 40 percent of our beef.

In some areas, the Ogallala has dropped precipitately. In the Texas Panhandle and in southeastern Kansas, declines exceed 100 feet. Declines of 10 to 50 feet have been measured in roughly half the aquifer underlying Texas, Oklahoma, Kansas, Colorado, and New Mexico. An engineering study, cited by Steve Coffel in his 1988 book *But Not a Drop to Drink*, suggests that by the year 2020 Texas will have lost 70 percent of its share of the Ogallala, New Mexico 60 percent, and Oklahoma 52 percent. The water shortage that follows will dry up 1.6 million acres of irrigated land in Kansas, 1.2 million in Texas, and at least a quarter million each in New Mexico, Oklahoma, and Colorado.

Agriculture is draining away surface waters as well. The nation's wetlands—including the grassy marshes typical of the plains—are disappearing at the rate of roughly 34 to 57 acres per hour, 24 hours per day. That is 300,000 to 500,000 acres yearly. Since the mid-1950s, we have lost 11 million acres of inland freshwater wetlands, particularly

Canada geese greet the misty dawn at a wetlands near Kearney, Nebraska. The vast plains offer food, shelter, and resting grounds for many migratory birds.

in the bottomland hardwoods of the lower Mississippi valley and in the prairie pothole region that includes the northern plains. About 80 percent of wetlands loss nationwide is due to agricultural drainage, often to create croplands for the growing of crops already produced in chronic surplus, crops that we then subsidize with price supports to the tune of several billion dollars yearly until 1985.

About 10 million wetland acres are acutely vulnerable to agricultural drainage. If present trends continue, we could lose up to 5 million more acres of wetlands by the end of the decade. It is more than a little ironic that rural people, who find their drinking-water sources increasingly laced with toxic contaminants, are destroying the very wetlands that could help filter toxins from agricultural runoff. Destruction of wetlands is also removing valuable habitat for birds, fish, amphibians, mammals, and many other animals. About one-half to two-thirds of all the ducks raised in North America come from the prairie pothole region. Some 30 percent of the plants and animals on the federal endangered species list require wetlands for survival.

Even rivers have been hard hit by irrigation plans. Irrigation accounts for about 80 percent of all U.S. water consumption, and the drawdown on rivers shows graphically how much this use taxes our water resources. For example, the Colorado River is so heavily exploited that in dry years virtually nothing is left to flow across the border into Mexico. The Platte River, which cuts horizontally across Nebraska from Colorado and Wyoming on its way to meet the Missouri, has lost about 70 percent of its flow to water projects. Agricultural interests have plans to sop up much of what remains, posing a threat to a half million sandhill cranes that stop along the river during spring migration—nearly the entire U.S. population—as well as millions of ducks and geese and several endangered species, notably the whooping crane and bald eagle.

Driven by external economic forces and by ill-conceived federal policies that reward excessive crop production and encourage the use of agrochemicals, American agriculture has grown dependent on chemical pesticides and fertilizers that have contaminated rivers, lakes, and groundwater, poisoned a wide variety of species, and threatened human health; it has pillaged the dry land and the wet land; it has caused the earth itself to billow off into the stratosphere; it has squeezed an array of species—from grizzlies to bison to pronghorn antelope—practically out of existence, and then it has drawn

Sandhill cranes coast down to their nightly roosts on the Platte River in Nebraska. The river's name comes from the French word meaning "flat and shallow." Numerous dams have reduced the river's water volume to a third or less of its historical flow.

upon the 98 percent of us who do not farm to pay for all this with tax-dollar subsidies. How did something as benign as farming, as gentle as feeding one another, get so out of hand? And what can we do about it?

Solutions

The first farmers were probably ants that raised fungi to feed to other species of captive insects upon whose secretions the ants fed. Doubtless ants have been farming for millions of years and will continue to do so—humanity willing—for millions more. Humans, on the other hand, started farming roughly yesterday, geologically speaking. It is hard to say just when we struck upon the plan of thrusting seed into soil and cultivating a crop. The only definite signs of it in archaeological digs are the remains of domesticated plants and animals. But long before these species were domesticated, our forebears doubtless cultivated the wild ancestors of these plants and animals. Thus it is impossible to tell whether some seeds stored by early peoples and found at archaeological sites were collected from cultivated plants or from plants that grew wild. The origins of agriculture lie in some lost era of human prehistory that began at least 12,000 years ago, somewhere in the Middle East and, perhaps simultaneously, in southeastern Asia.

The idea of cultivation spread rapidly, and the world has never been

Plains tickseed wildflowers explode on the Oklahoma prairie. During dry periods, the expansive grasslands may seem monotonous, but closer observation reveals a wealth of species and minihabitats for wildlife.

the same. Agriculturalists helped create deserts in parts of Asia and Africa, using the ever-efficient desert-making tool of overgrazing. They cut down the looming forests of Europe and, much later, North America, getting all those junky trees out of the way of crops. They made the world safe for corn and wheat.

They were pretty cavalier in their approach to the natural environment, cutting ancient trees at will and draining wetlands. Right up until the early twentieth century, their perspective on these things was different from ours today, or at least different from that of some of us now. Natural areas were not something to protect. They were to be subdued. God had put them there for that purpose. Clearing the land was an allegory for the taming of the human soul, for coming out of the wilderness and into grace. If for some individuals that was not a good enough reason for turning wild lands into farmlands, then making money was. Any fool could see that you could make more money destroying wilderness than preserving it. Wilderness was a worthless commodity. There was too much of it. In an era in which muscles, not motors, powered most modes of transportation—an era that lasted roughly from the dawn of humanity until about 1910—the world seemed too big to succumb to human schemes. There would always be more open land and more wildlife, but we could never be sure that there was quite enough farmland or ranch land or city.

We can be sure of it now, though. We could call it quits right this moment in the game of global environmental conquest and still see that we have massacred the other team. Only a few areas of the world are relatively untouched by humanity, and they are vanishing fast. The regions that we are putting to heavy use are beleaguered and failing. For our own good we should now give the globe a helping hand, and in few places is this as true as in our agricultural lands.

Some remedies are already in place. The farm bills of 1985 and 1990 included measures that attacked the problems of wetlands drainage and the plowing of marginal lands that easily erode. The so-called swampbuster and sodbuster amendments take away all federal subsidies from farmers who convert wetlands and fragile lands to crops. The Tax Reform Act of 1986 removed deductions for expenses incurred by farmers preparing land for center-pivots or converting wetlands to croplands. It also eliminated capital-gains protections for farmers who convert erodible lands or wetlands to crops. The 1990 law also requires farmers to keep records on their use of pesticides governed by federal law.

The farm bills have also brought some positive actions. The 1985 law created the Conservation Reserve Program which, in four years, removed 34 million acres of excessively eroding cropland from production and initiated the planting of grasses and/or trees on the removed lands. Thus, some 10 percent of the nation's croplands are being restored to relatively healthy natural conditions after uses that threatened to doom them not only as agricultural land but as wildlife habitat. The 1990 bill led to a $46.4 million appropriation for permanent easements on some 50,000 acres of wetlands that had been converted for farming. The bill's Water Quality Incentive Program received $6.75 million for payments to farmers working to prevent pollution of groundwater and surface waters.

Nevertheless, abuses continue. The swampbuster program, for example, only works against farmers who participate in a subsidy program and who convert a wetland *and* plant it with crops. In addition, the local committees charged with implementing the swampbuster measures are composed of farmers who often ignore, or liberally interpret, the measures. Violators are routinely excused from penalties.

Correcting the many problems that surround farming, both in the Great Plains and elsewhere, will require a well-integrated program that seeks to ensure the integrity of the land, human health, and farm profits. Such a plan was outlined in *Farm Bill 1990: Agenda for the Environment and Consumers*, a special publication of 11 environmental groups that included the National Audubon Society, the Soil and Water Conservation Society, the Sierra Club, and the National Resources Defense Council.

The plan suggests development of a national policy to control the sources of water contamination stemming from agriculture. The policy, says the report, "should be made a formal, central feature of U.S. agricultural and environmental policy." This "involves cutting back on the use of pesticides and other potential environmental contaminants, or reducing the generation of other 'waste' through modifications of farming systems and practices." This policy would include a 10-year period during which farmers could identify and implement cost-effective means to reduce the use and generation of pollutants. Farmers should be required to keep records of annual pesticide use so that federal and state agencies can monitor trends in use and provide advice on ways to reduce the reliance on pesticides. Similar initiatives should be created for fertilizer use.

To help farmers implement these measures, the groups call for development of an Environmental Stewardship Program that would provide financial assistance to farmers attempting to protect wetlands and erodible lands and to reduce environmental contamination from pesticides, fertilizers, sediment, and animal wastes.

Cutting back on the use of fertilizers and pesticides can save money for farmers. For example, a recently developed soil test may help reduce the application of nitrogen fertilizer on corn by 30 percent in Iowa, resulting in a yearly savings of $100 million. In addition, ending the reliance on chemicals would force farmers to resort to other means of maintaining soil productivity and controlling pests, means that in the long run are more effective than the chemical quick fix. For example, rotating crops not only restores soil nutrients but has other benefits as well. Studies in the Midwest indicate that alternating corn with crops such as clover and alfalfa increases crop yields. Rotating crops also disrupts pest life cycles, diminishing pest numbers inflated by the easy pickings of year-to-year single cropping. Rotating corn with other crops eliminates the need to use a soil insecticide to combat corn rootworm. Also, getting away from pesticides has other advantages. Insecticides in some areas have wiped out honeybees needed for crop pollination. Farmers have actually had to restock bees in areas where they wiped out the insects with pesticides. Insecticides have even increased insect depredations in some areas by killing off species that prey on pest insects while leaving the pests relatively unharmed.

The federal government also needs to enforce existing laws more consistently and vigorously. Swampbuster penalties should be applied whether or not the violator plants crops on the drained land. The federal government should also initiate programs that teach and encourage the use of organic farming techniques that focus on natural fertilizers, such as manure and compost, on plowing techniques that minimize erosion, and on nonchemical pest controls. Countless studies have shown that "organic" techniques produce crops at less cost per acre to the farmer, because of savings in chemical additives, and create a more productive soil with a higher capacity for water retention and higher levels of nutrients.

At present, American agriculture, particularly in the Great Plains, is fighting the ecological flow. It is draining away the water that makes irrigation possible and poisoning the wellsprings of its own life. Long ago, federal treaties promised Native Americans rights to various lands for "as long as the grass shall grow, and the rivers shall flow." Today one could almost justify the breaking of those long-lost treaties: In the Great Plains, native grasses have ceased to grow in most areas, limited primarily to patches in cemeteries and along roads; most rivers no longer flow freely, for they are harnessed by dams. But until our farmers heed the ecological needs of the land they cultivate, they will enjoy at best only a fading illusion of prosperity. Frederick Kirschenmann put it succinctly in an article for the Winter 1988 issue of the *American Journal of Alternative Agriculture:*

> More economists are recognizing that the cheap food produced by fewer and fewer farmers within our conventional agricultural system is not the good deal we thought it was—because not all of the costs have been factored in.
>
> We haven't factored in the cleanup costs of decontaminating our groundwater, for example. We haven't factored in the costs of restoring the soil which has been eroded and salinized at unprecedented rates. We haven't factored in the taxes used to subsidize the oil and water required to produce crops conventionally. We haven't factored in the costs of social destabilization caused when large numbers of farmers were lost from rural communities. Nor have we factored in our increased health care costs, clearly related to conventional agricultural practices. If we were to factor in all of these costs, it would become painfully clear that we cannot prosper economically, even in the short term, without attending to ecology.

OUR FATAL SHORES

In the second decade of the nineteenth century, Lord Byron, scribbling the lengthy work that would make him famous—*Childe Harold's Pilgrimage*—put pen to paper and wrote of the sea. The sea was something close to Byron's heart. He was an avid and powerful swimmer who, despite a clubfoot, succeeded in swimming long distances in ocean surf. No stranger to traveling under sail, he was a familiar of the seas, which in their vast, heaving mystery seemed to him incapable of being scarred by humankind:

> Roll on, thou deep and dark blue ocean, roll!
>   Ten thousand fleets sweep over thee in vain;
> Man marks the earth with ruin, his control
>   Stops with the shore; upon the watery plain
>   The wrecks are all thy deed, nor doth remain
> A shadow of man's ravage, save his own,
>   When for a moment, like a drop of rain,
> He sinks into thy depths with bubbling groan,
> Without a grave, unknelled, uncoffined, and unknown.
>
> His steps are not upon thy paths, thy fields
>   Are not a spoil for him,—thou dost arise
> And shake him from thee; the vile strength he wields
>   For earth's destruction thou dost all despise,
>   Spurning him from thy bosom to the skies.

Byron wrote of the purity and power of the sea with a belief in its invulnerability that we in the twentieth century can no longer enjoy. Human society has invaded the oceans, saturating them with pollutants and crowding them with so much debris that now the seas are regurgitating wastes and poisons on shorelines throughout the world. What once seemed infinite in its ability to absorb the shocks and insults of human wastes is showing all the signs of overload and collapse.

In the United States, organized dumping of wastes into the ocean began during the colonial era. Philadelphia established our first municipal waste-collection system: Slaves laden with trash waded into the Delaware River to dump their loads into the prevailing current.

Of course, thanks to modern technology and innovation, we have taken the dumping of wastes to new heights and depths. Every year, barges carrying sewage sludge from New York and New Jersey dump some 9 million tons of the stuff at a disposal site 106 miles off the Jersey Shore. Private organizations contribute their share, too. Ciba-Geigy dumps 4 million gallons of contaminated wastewater off New Jersey every day, and Allied Chemical infuses the New York Bight—an oceanic triangle bordered by New Jersey and Long Island—with 50,000 tons of acid wastes yearly. Each year from 1975 to 1985 an average of about 600,000 barrels of oil were accidentally spilled into the ocean from tankers and other sources, according to the U.S. Na-

The sun scatters its late-afternoon light on New Jersey coastal waters, turning rotting pilings into modern art.

tional Research Council. This was a relatively minor source of oil pollutants, however. An additional 21 million barrels seeped into the sea annually from such sources as street runoff, effluent from industrial facilities, and the flushing of tanks on ships.

In all, about 16 trillion gallons of sewage and industrial waste, much of it tainted by chemical toxins, are released yearly into U.S. rivers and coastal waters. The effects of such dumping can be seen all along our coasts and waterways. Signs erected along some 100 miles of the heavily industrialized St. Lawrence River warn that shellfish there are no longer safe to eat, even in relatively pristine areas remote from industry: Pollutants travel far on river currents. New York health officials have warned against eating bass, carp, sunfish, catfish, and walleye from the Hudson River. The striped bass, which spawns in the river and New York Harbor, accounted for a catch of nearly 15 million pounds in 1973, making it an important commercial species. Now stripers are so laden with toxins, notably cancer-causing polychlorinated biphenyls (PCBs), that they are banned from commercial seafood markets. Roughly 40 percent of shellfish beds across the United States have been closed because of pollutants. Flounder from Boston Harbor have been rendered inedible by toxins. Indeed, all fishing has been banned in Boston Harbor.

A study of creatures most likely to show signs of toxic pollution—shellfish and bottom-dwelling fish such as flounder—was recently

Sun and surf worshipers invade beaches such as this one at Brigantine, New Jersey, vying for position when temperatures climb. A flood of floating syringes and other medical wastes has been brought under control, but some experts maintain that invisible health hazards still linger.

A shipping graveyard graces the waters off Staten Island, New York. In addition to being eyesores, such sources of rust and paint peelings add tons of pollution to our waters every day.

conducted by the National Oceanic and Atmospheric Administration at 200 sites around the nation. It revealed, not surprisingly, that the areas bearing the greatest amount of pollution are those with dense populations and a long history of industrial discharge. Among the worst are on the East Coast, specifically such areas as Boston Harbor. But also high on the list were San Diego Harbor and Washington State's Puget Sound. Seattle's Elliott Bay, in the heart of Puget Sound, is contaminated with PCBs, copper, lead, arsenic, zinc, and cadmium. Some parts of the sound feature signs in seven languages—English, Spanish, Vietnamese, Cambodian, Laotian, Chinese, and Korean—warning that crabs, shellfish, and bottom fish may be unsafe to eat.

San Francisco Bay, which has lost 90 percent of its wetlands to development—wetlands that could have helped purify wastewater dumped into the bay—is contaminated with copper, nickel, mercury, cadmium, silver, selenium, and other heavy metals found in industrial wastes. PCBs have impaired reproduction in starry flounders and in black-crowned night herons that nest on Bair Island, part of a wildlife refuge. PCB levels in the tissues of the bay's harbor seals match those that caused reproductive problems in gray seals in the Baltic Sea and in ringed seals from the Gulf of Bothnia, which lies between Finland and Sweden.

Toxins alone are not the problem. Certain plant nutrients, such as those found in sewage dumped at sea and in fertilizers that flow daily by the ton into coastal waters as runoff carried in rivers, may pose another set of dangers. These nutrients might cause naturally occur-

Signs at Eagle Harbor, across from Seattle, Washington, offer a seven-language warning against consumption of certain seafoods. Such reminders of our failure to control pollution are now commonplace on both coasts.

ring species of algae—microscopic plants—to reproduce much more rapidly than usual, crowding the surface in such teeming numbers that the water takes on the color of the algae. Such runaway growth is called *algal bloom*. The most famous type is caused by red algae, so algal blooms are generally called *red tides*. However, blooms may be yellow, brown, green, or other colors.

Red tides pose several threats to the marine ecosystem. Some algae produce toxins that kill fish and cause human illness. In August 1988, a red tide off Naples, Florida, crippled local fishing and tourism businesses and led to 41 reported cases of respiratory, gastrointestinal, and neurological illness among people who spent time along Naples' shores. Even a nontoxic bloom can damage an ecosystem. The algae can become so dense that they block sunlight from the sea, starving or forcing away many of the surface-dwelling microorganisms that serve as the foundation of the marine food chain. Moreover, when algae die, they sink to the bottom, where their decomposition uses up oxygen dissolved in the water, leading to suffocation of fish and other animals. Long Island's Peconic Bay has lost its $1.8 billion scallop industry because a persistent algal bloom has snuffed out the scallops' food sources. The algae also seem to have wiped out beds of marine grasses and to have reduced conch populations and the numbers of small fish fed upon by such commercially important species as mackerel, flounder, and bluefish.

Another type of human-caused marine degradation is solid debris: abandoned fish nets, plastic bottles, packing pellets, cans, and myriad other objects. Many of these objects end up in the ocean after being dumped from boats, falling from garbage barges, or blowing from landfills. Some float at sea for a long time. An aluminum can may take 500 years to degrade. Glass may last a million years. Some plastics will never break down. When we are little more than fossilized remains, we can rest assured that plastics, a hallmark of the twentieth century, will still plague the world in much their original condition.

Plastics will become an increasingly difficult problem in the future. From 1970 to 1985, demand boosted U.S. production of plastic from 19 billion to 48 billion pounds yearly, according to the Society of the Plastics Industry, Inc. The society predicts that annual demand will increase to 76 billion pounds by 2000.

Plastic is by no means our only lasting gift to the sea. A study by the National Academy of Sciences in 1975—which remains the most comprehensive worldwide report on marine litter—indicated that at that time glass and metal items, as well as tar balls from oil production, occurred in the ocean in greater number and volume than did plastic waste. Of an estimated 6.4 million metric tons of garbage reaching the ocean in any given year in the mid-1970s, plastics composed only 45,000 metric tons. More recent, if limited, studies suggest that plas-

Environmental scientists for the Environmental Protection Agency take a water sample from New York Harbor. The water's temperature and acidity are measured from the helicopter, and the sample is analyzed at the lab in Edison, New Jersey.

tics may have become a bigger problem since the mid-1970s. On Sable Island, about 100 miles off Nova Scotia, a three-year volunteer study in the 1980s examined the amount of garbage reaching six sites, each about a third of a mile long, on the island's north side, where most debris accumulated. In April, all debris was removed from the study sites, and then new debris was collected about every 40 days through November. In three years, 11,183 items were picked up at the study sites, an average of nearly 2,000 items per mile. Plastics composed about 94 percent of the total, glass about 5 percent, and metal less than 1 percent. About 35 percent of the debris consisted of discarded fishing gear.

Such debris is deadly to wildlife. Discarded fishing gear—amounting to hundreds of thousands of tons annually—entangles and kills a variety of marine species, including fish, dolphins, seals, and birds. Entanglement in trawl webbing and packing bands may be the principal cause of a steady decline in the northern fur seal population in Alaska's Pribilof Islands since the mid-1970s. Recent studies in Alaska indicate that every year as many as 30,000 northern fur seals—out of a population of about 827,000, representing some 70 percent of the world population—die from entanglement. The Alaska population numbers less than half what it did 30 years ago and is declining 4 to 8 percent yearly.

Ingestion of plastic bags and tar balls may be a major threat to sea turtle survival, according to the late turtle expert Archie Carr. Little data exist for determining how widespread this problem is and how it is affecting sea turtle populations, but a researcher on Long Island, New York, found that 11 of 15 dead leatherback sea turtles had eaten plastic bags or monofilament-plastic fishing line. One 12-pound juvenile hawksbill turtle found in Hawaii had eaten nearly 2 pounds of plastic, including a plastic bag, golf tee, monofilament fishing line, a plastic flower, a comb, and dozens of small miscellaneous pieces of plastic. Such items can kill an animal by blocking its intestine. Eating plastic packing pellets also may alter an animal's buoyancy, affecting its ability to dive for food.

Byron wrote, *"Man marks the earth with ruin, his control / Stops with the shore."* But clearly this illusion has come to an end. From Boston Harbor to Puget Sound, coastal waters are sullied under the onslaught of human waste and degradation. In many areas, pollutants pose serious threats to human health. What we have done to the oceans is a sign of the vast extent to which we have threatened the health and integrity of the global habitat that sustains our lives.

Boston Harbor: This Ain't No Tea Party

In Chelsea County, east of Boston, housed in a turn-of-the-century red-brick building on the edge of Boston Harbor, is a municipal sewage-pumping station designed to remove bricks, two-by-fours, and other large objects from sewage waters. Unfortunately, it was designed to do this in about 1890. In its doddering old age, the pump is a symbol of much of what is wrong with sewage treatment in the Boston area and an emblem of what lies at the heart of Boston Harbor's pollution problems.

Not that it isn't a beautiful piece of equipment. A sewage pump of modern design would never display such elegance. It is an object at which to marvel. It has been pumping sewage for the city of Boston since 1893, but it is still gleaming and bright, painted lipstick red and trimmed with glistening brass. The building that it dominates is itself a relic of the past, a sort of brick time warp. The original woodwork is still polished and clean. Old-time equipment—made in an age that never knew plastics—lurks in many corners, including a meter several feet tall that indicates water depth beneath the building.

But the star of the pump station is clearly that dazzlingly red pump. Its operation requires three men performing eight hours of daily maintenance, and when it is running—which is infrequently, since it is used primarily as backup during times of heavy sewage loads—it requires constant monitoring of oil levels to keep it lubricated. Some of the workers are third-generation attendants to The Machine, following in the footsteps of their fathers and grandfathers.

There is something beautiful and quaint and pleasant about this place. It is truly a living remnant of a more leisurely time, right down to the white cat that lives there, rubbing the legs of visitors and curling up in cozy corners. But unfortunately, many of the antiquated aspects of Boston's municipal sewage-treatment system—though equally old or older—are not so quaint, far less beautiful, and far from benign in their effects.

Consider the city's 5,000 miles of pipes that carry sewage from some 43 surrounding communities to the Boston treatment plants. Some of these pipes have been in service for more than a century. Some of them are made of wood, and leak badly. Worse, the oldest pipes carry not only sewage but street runoff, which means that during heavy rains they overflow, and raw, untreated sewage pours into Boston Harbor and the rivers that feed into it. This explains why, after a big storm, you can find the Charles River afloat with condoms, tampon applicators, orange peels, bits of plastic, and other debris. This is a violation of the federal Clean Water Act, passed in 1977, which is why Boston was forced to undertake construction of the largest public works project ever proposed in New England—a $6 billion wastewater- and sludge-treatment plant that will bring Boston's sewage treatment up to the level standard in such places as Chicago and Milwaukee since the 1920s.

Being so far behind the times has put Boston in an unenviable position for two reasons. One, it sits on the edge of one of the nation's most seriously polluted bodies of water. Fishing is banned there, because fish pulled from Boston Harbor are too toxic to eat. Two, it is going to cost the city a lot of money to build the plant. Consequently, the city—which, in a burst of local-rights fervor, some 10 years ago turned down a federal offer to cover 90 percent of the cost of improving the municipal sewage-treatment system—is faced with increasing fourfold its sewage treatment bills to residents. By the year 2000, the yearly fee will have risen from about $90 per household to about $1,500.

At present, Boston dumps some 500 million gallons of treated sewage into the harbor every day. This volume, according to a press packet prepared by the Massachusetts Water Resources Authority, is equal to the amount of water that flows into the ocean daily from the Charles, Mystic, and Neponset rivers combined. The sewage includes industrial and household waste gathered from 43 communities surrounding Boston and carried by pipeline to two treatment plants—one at Nut Island on the western side of Boston Harbor, the other at Deer Island, farther east.

Deer Island will be the site of the new treatment plant. It was once, presumably, home to deer, but it has long since fallen into harder times. At one time, immigrants were quarantined on Deer Island. It has been the site of a prison since the nineteenth century. It boasts a cemetery in which were buried long ago people who died of contagious diseases. In the colonial era it had a fort at one end from which a net extended across the harbor to another island so that the net could be used to block enemy ships.

The biggest feature on the island today, aside from the prison, is the sewage-treatment plant. Sewage arriving there is first run through screens to filter out objects more than an inch wide. The accumulated screenings are loaded into trucks and shipped off to Buffalo, New York, where they end up in a landfill.

Coarsely filtered sewage then flows into a grit chamber, where sand and mud settle out. The grit is then dumped in a landfill. The remaining sewage flows to sediment tanks to finish a process called *primary*

◄

A worker shoots a cleansing stream of hot water over machinery at the Deer Island sewage-treatment facility. A new treatment center carries a $6 billion price tag and will mean a 1,500-percent increase in sewage fees by the year 2000.

treatment. In the tanks, few toxic chemicals are removed, but about 40 percent of the human waste settles out to form mudlike sludge. Floating objects, such as plastic and fat, constitute what those in the sewage-treatment trade call *scum*. The new plant will be able to sort out the plastics, so in the future the remaining nonplastic scum will be dumped at sea via a tunnel bored some nine miles into Massachusetts Bay. The tunnel, now under construction as part of the new plant, is being cut through bedrock some 400 feet underground. It will replace one that goes only a mile into the bay. Presently, scum is treated to kill bacteria and then is taken to a landfill. If the unseparated plastics and other solids were dumped into the harbor, the floating debris would come back to shore.

The wastewater from the sediment tanks next enters *secondary treatment*. At this stage, oxygen is added to speed the growth of microorganisms that consume up to another 50 percent of human wastes and up to 70 percent of toxic chemicals. The microbes settle to the bottom of the tank as sludge. The wastewater is treated with heavy doses of chlorine and dumped into the harbor to the tune of some 300 million gallons daily. This poses two problems for the harbor, neither of which has been thoroughly examined. First, the chlorine is toxic and may harm plant and animal life in the immediate area of the dumping. Second, only about 30 percent of organic solids have been taken from the water, according to Diane Jurgen of the Massachusetts Water Resources Authority's public relations staff. Remaining organic matter may increase nutrient levels in the harbor, leading potentially to algal blooms.

The sludge from both the primary and secondary treatment tanks is put into a digester to reduce their volume through loss of moisture and gas. The digester produces methane that, Jurgen says, could be used to power the sewage plant's pumping stations, but that at present is simply burned. Indeed, flames from the process flicker above four smokestacks in the roof of the plant. The remaining sludge is further dried and processed to form pellets that can be used as crude fertilizer. The new plant will be able to process sludge into a more refined fertilizer so that it can be sold for use even in golf courses and citrus groves. If the water authority cannot market the sludge, which currently does not meet federal health standards after processing, its landfills will fill within 25 years.

The Boston sewage-treatment system is essentially a nineteenth-century artifact with a few technological updates grafted onto it. No doubt many of Boston Harbor's troubles can be traced to the treatment system. The Massachusetts Water Resources Authority boasts that the current treatment system removes 90 percent of sewage wastes and up to 70 percent of its toxins, both household and industrial. Looked at from another perspective, the 300 million gallons of wastewater dumped into the harbor each day still bear 10 percent of their human wastes and 30 percent of their toxic load.

This level of waste effusion, combined with other waste sources such as street runoff and industrial effluents, has contributed to making Boston Harbor one of the nation's most polluted bodies of water. The harbor's fishing industry is dead. Commercial fishermen are

◀

Teachers on the "Envirolab" research boat pull in the Peterson Bottom Grab, which is full of bottom sediments from Dorchester Bay, Massachusetts. The group will examine the sample and analyze its properties.

forced to go 75 to 150 miles off-shore to be certain of catching uncontaminated fish. At least one local seafood restaurant chain, Legal Seafood, has felt pressed to create its own laboratory for testing the bacterial content of the shellfish it sells. The restaurant's headquarters receives its shellfish shipments at about 10 A.M. and the incoming shellfish are quarantined for 24 hours, says Legal Seafood owner Roger Berkowitz. During that 24-hour period, tests are run to determine the bacterial levels of the sample shellfish. Legal Seafood runs more than 2,000 shellfish tests yearly. The lab is also developing the capability to test for toxins, a process that with current methods takes up to 10 days to complete. Legal Seafood is attempting to cut that to 24 hours.

Virtually all of the fish sold by Berkowitz comes from some 80 miles or more offshore, in hopes of ensuring its purity. Almost all harbors, he says, are too heavily polluted to provide food for his restaurants. He echoes the concern of many who feel the loss of the harbors as sources of food when he says: "Well, it's sort of like having a precious commodity like gold in your fingers—in your hands—and letting it slip through your fingers and knowing that unless someone stops it quickly it's gone for good. If we don't wake up to the fact that we're polluting our waters we could be in a lot of trouble, because it's too good to waste, too good to waste."

Boston does seem to be waking up to the problem. Says Paul Levy, director of the Massachusetts Water Quality Authority, "I'm cautiously optimistic about the Boston situation. . . . Boston has awakened. It's starting to fix its problems and make a lot of investments."

COASTAL POLLUTION

THE ISSUE The coastal regions of the United States are heavily polluted with wastes from urban disposal systems, raw sewage, industrial toxins, airborne toxins washed in by rain, farm runoff rich in fertilizers and pesticides, and urban runoff contaminated by petroleum products, lawn fertilizers, and pesticides.

THE CAUSE Antiquated city sewage systems overflow during heavy rains, permitting untreated sewage, including human feces, to run into harbors and streams. Rainfall on farms washes off the many toxic chemical compounds used by farmers, such as pesticides, as well as fertilizers whose nutrients cause lakes, rivers, and harbors to become choked with unnaturally high concentrations of algae, often with toxic effects on aquatic wildlife. Rain runoff in cities carries lawn chemicals as well as oil and gasoline from streets into nearby streams, lakes, and coastal waters. Air pollution, such as acid compounds and toxins generated by motor vehicles, also reaches the seas in rainfall. The dumping of industrial, urban, and military wastes directly into the oceans causes problems when containers leak or when materials are simply poured into the sea. Wastes dumped into rivers also reach the ocean eventually.

EFFECTS ON WILDLIFE Some experts believe that toxic pollutants were behind the 1987 die-off of half the Atlantic coastal population of the bottlenose dolphin. Some fish populations and the birds that feed on them are probably at risk, as well as a variety of marine mammals, including seals, sea lions, whales, and dolphins. Sea turtles, too, have shown some signs of possible contamination. For example, 80 percent of the sea turtles examined by biologists at Florida's Indian River, which is a coastal bay, show tumorous growths probably related to toxic contamination.

EFFECTS ON HUMANS Some medical professionals believe that the contamination of beaches is leading to an increased incidence of various types of illnesses in children. Eating contaminated fish and shellfish is a major health threat as well. Contaminated fish populations also lead to loss of jobs and millions of dollars in income in the fishing industries.

EFFECTS ON HABITAT Oceans and beaches are increasingly contaminated with bacteria from human wastes, causing beach closures. Fish, marine mammals, and other sea life are faced with loss of oxygen and of food supplies, with corresponding declines in population.

Of course, Bostonians are not alone in coming alive to the dangers that lurk on their shores.

The Keeper of Long Island Sound

Every day, 40 sewage-treatment plants in New York State and Connecticut pour a billion gallons of wastewater into Long Island Sound. An average of about 2 million gallons of raw sewage hits the sound daily, particularly on days when more than a quarter inch of rain falls, overloading the sewage system to the point that raw sewage simply overflows municipal pipes and floods into the sound. That is roughly 800,000 gallons of sewage water for every one of the sound's 1,300 square miles of surface area. This wastewater is rich in the nutrients typically found in agricultural fertilizers, such as nitrogen. Indeed, some 91,000 tons of nitrogen enters the sound each year, more than half of it from human activities such as sewage treatment, street runoff, and air pollution. The nutrients hasten growth and reproduction in microscopic algae, stimulating algal blooms. The blooms in turn lead, ultimately, to the loss of dissolved oxygen in the water. During especially warm weather, oxygen levels can drop over large areas of the sound. In the unusually hot summer of 1987, dissolved oxygen fell to hazardous levels in roughly the western third of the sound. Some 800,000 fish died, putting a heavy burden on commercial fishermen who depend on the fish for their livelihood. Similar conditions have occurred in the area every year since. In 1990, dissolved oxygen dropped almost to zero in the sound from the Bronx to New Haven, an area covering some 500 square miles. Moreover, contamination by raw sewage has forced states to close commercial shellfish beds, cost-

Shellfish are laid out for testing at the Legal Seafoods lab. All catches are quarantined until they receive the lab technician's seal of approval. Restaurant patrons can sleep better at night, knowing that the Legal laboratory tests all batches of incoming merchandise.

ing Connecticut alone more than $15 million in lost annual revenues. In all, nearly 146,000 acres of potential shellfish beds in the sound are closed or restricted because of contamination by bacteria and other pathogens.

The sound is the victim of overdevelopment. The sound and regions within 50 miles of it are home to 10 percent of all U.S. residents. During the 1950s, 1960s, and 1970s, wetlands that edged large portions of the sound and helped filter pollutants out of runoff water were drained to create space for housing and other building projects. Roughly half of all the sound's wetlands have been lost since colonial times. Since 1960, development reached more than two-thirds of the upland areas around the western half of the sound, more land than was developed in the previous 300 years. Such extensive development in turn increased the amount of urban runoff—replete with pollutants such as oil and gasoline that wash off of streets during rains—that reaches and contaminates the sound. Industry along the sound also dumped wastewater into bays, creating regions of highly polluted bottom sediments. Industrial chemical toxins entered the food chain, forcing the shutdown of commercial shellfishing in various hotspots.

Perhaps only a tiny percentage of the 5 million people living on the shores of Long Island Sound have recognized the problems that lie on their doorstep. But many of those who are aware have banded together for action. One such group is the Long Island Sound Watershed Alliance, created in January 1991 at the culminating conference of the National Audubon Society's year-long Listen to the Sound Citizen's Hearing Campaign. As part of the campaign, Audubon—with the cooperation of more than 175 other organizations—slated 15 public hearings in the Long Island Sound area. The hearings were pat-

Fresh from the boat and delivered to historic Boston pier, fish are sorted and packed into bins before being weighed and sold to awaiting buyers. Fishermen sail 75 to 150 miles offshore in hope of catching uncontaminated fish.

terned after traditional New England town meetings in which participants were limited to 5-minute oral presentations of their viewpoints. More than 1,500 people aired their grievances about the declining quality of the sound. The Watershed Alliance, a direct result of citizen concern, is an information- and resource-sharing network for organizations working to protect and restore the biological integrity of the sound and its surrounding watershed. The alliance included more than 50 organizations by December 1991 and is still growing. The alliance is working at the local level to identify problems and solutions.

Of the many people who have worked through groups such as the Long Island Sound Watershed Alliance, few have garnered as much press as Terry Backer, a third-generation fisherman. His father and his grandfather before him also made their living from the shellfish beds of Long Island Sound. Backer has fished off California, Oregon, Washington, and Alaska. He has fished the Atlantic, and he has set lobster traps and scoured oyster beds in Long Island Sound. He is a bearded, burly man who downs a dozen cups of coffee a day and peppers his speech with epithets. He is, above all, the keeper of the sound, a position he earned through his endless fights to save Long Island Sound from human irresponsibility.

Not that Terry Backer would call himself an environmentalist. No way. To him, environmentalists are cheese-and-cracker-snapping, tree-hugging hobbyists with too much time on their hands. No, despite his concern with the integrity of the sound and all the lawsuits in which it has involved him and all the municipalities he has sued to force them into wiser disposal of wastes, he calls himself a fisherman and leaves it at that.

As well he may, since it was as a fisherman that he first came in contact with the sea he loves and the problems that plague it. As a little boy only 8 years old, he would complain when his father threw cigarette butts over the side of their fishing boat. As an adult, he was appalled one day when he cruised through Norwalk Harbor, an inlet of Long Island Sound, and found it afloat with human waste and excrement. But his concern about the environmental impacts of human activities was shaped not just by what he saw before him, but by what he sensed lay in the years ahead. This is clear in the way Backer responds when asked why he became involved in environmental issues: "Well, my first impulse is to tell you that I love the water and I love the sound, but a lot of people do, and I don't think that's it. When my son Jacob was born I looked around and saw that each generation was leaving less behind it, and I felt it was my responsibility to put up a fight for him. Maybe it has more to do with my kids than it has to do with the sound. The sound has certainly provided for me and my dad and my family for a long time. I mean we owe it, we owe it something."

Backer's commitment turned to action by an indirect route. In spring of 1986, he cruised from Norwalk, Connecticut, to Nyack, New York, a town on the Hudson River, to buy shad with which to bait his lobster traps. He learned from a fellow fisherman there that the Hudson was afflicted with some serious problems. River pollution had be-

come so bad in the 1970s that commercial fish species had been contaminated. Striped bass, catfish, and eels were pulled from the market in the name of public safety. This cut into the incomes of lifelong fishermen so drastically that they banded together to form the Hudson River Fishermen's Association. Under this aegis, they engaged in battle with various polluters, municipalities, and regulatory agencies. In 1984 one of the fishermen, John Cronin, was named riverkeeper by New York State and assigned to patrol the Hudson in search of environmental law breakers.

The idea of the riverkeeper came from an old English tradition of assigning special guards to monitor the health and use of streams belonging to the crown. As a riverkeeper, Cronin kept up his end of the tradition. Even before he was officially named riverkeeper, he won a lawsuit against Exxon—the petroleum company that would one day spill 11 million gallons of oil into Alaska's Prince William Sound—after he caught an Exxon tanker miles up the Hudson flushing its ballast tanks and dumping toxic benzene, toluene, and other chemicals in order to take on water to sell and use at Aruba, a Carribean island on which Exxon has a petroleum refinery. Both the flushing and the taking of water were illegal. Exxon ended up paying New York state $1.5 million in fines and provided $250,000 to establish the Hudson Riverkeeper Fund. Cronin was then given his title. Once established as riverkeeper, he also won a famous case against New York's Westchester County when he caught workers pumping toxic wastewater from a county dump into the Hudson on Christmas Day. When Cronin sued, the court ordered the dump closed and required the county to develop a new waste-management plan.

When in Nyack, Backer looked up Cronin and learned how to bring the environmental fight to Long Island Sound. Backer started by helping to form and lead the Connecticut Coastal Fishermen's Association, an alliance that grew to include lobstermen, boat owners, and even swimmers. In 1986, the association entered its first legal battle by suing the cities of Norwalk and Bridgeport when sewage emanating from their systems forced the closing of Norwalk's oyster beds. The association charged the cities with 5,400 violations of the federal Clean Water Act. A year later, Norwalk settled out of court and agreed to replace and repair the sewage-treatment equipment that had caused the overflow of raw sewage into the harbor. Norwalk also coughed up $172,500, half of it earmarked for cleaning up Norwalk Harbor. The other half was used to create the soundkeeper position. Backer took the $400-per-week job, which entails educating the public, lobbying for more effective pollution management programs, and reporting instances of illegal sewage disposal and industrial pollution.

In 1988, Backer and the association struck again, suing Greenwich, New Haven, and Stratford. Though Bridgeport has yet to settle on the case that began in 1986, the results in the other cases have been stunning. Stratford signed an out-of-court agreement in early 1990 and is updating its treatment plant. Greenwich is still negotiating with the association, but has initiated plans for a $40 million addition to its plant. New Haven has paid nearly $200,000 in fines and has begun to upgrade its old equipment and hire more staff. Even Bridgeport is hiring more staff and repairing some equipment.

◄

Long Island Soundkeeper Terry Backer is one of the waterway's watchdogs. The son of the son of a waterman, Backer tests outflows and monitors shipping practices to help keep the sound healthy.

Backer cruises the sound constantly in his boat, checking sewage outfalls and taking water samples to determine if cities and industries are dumping toxins illegally. He is part of a growing cadre of citizen-activists across the nation who have taken upon themselves responsibility for the integrity of the nation's shores. He is matched by keepers in such diverse places as the Delaware River and San Francisco Bay. But Backer and others point out that official and semiofficial keepers cannot alone solve our pollution problems. They must be joined by 5 million, 10 million, or 50 million other citizens willing to keep their eyes open for pollution problems. Every individual who sails Long Island Sound, for instance, can actively look for such violations whenever out on a cruise.

Backer points out that not all the problems that plague the sound stem from the actions of municipalities and industry. He has also witnessed the cumulative impact of a burgeoning population. One example is pollution with petroleum products. "When you think about oil, you think about the Amoco *Cadiz*, you think about the Exxon *Valdez*, you think about the big, big oil spills," says Backer, referring to the two American tankers that made history by causing massive spills. "But we have a continual oil spill on our coast. It happens every day. We have millions and millions of cars going up and down the road leaking hydrocarbon products—oil, grease, gasoline, ethylene-glycol

Heavy shipping is vital to Seattle-Tacoma economies but brings with it uncounted sources of pollutants. Ship-sandblasting operations and metal scrappers are being monitored to control the amount of paint chips and dust entering Puget Sound. The sound's waters may look pristine, but below decks lurks a toxic monster.

antifreeze. All this stuff goes down to the road and you know it doesn't disappear. And when it rains it comes down these storm drains and into the sound. We're the top of the food chain. We're the people who get it all in the end."

The sheer mass of people living in the northeastern United States has led to widespread environmental degradation. As Backer puts it, "We've gotten used to saying we can't eat from here any more because it's polluted, we can't swim here because it's polluted, we can't take the kids to play by the marsh area because it's gone, and we've gotten so used to just moving over that we accept anything that comes down the pike."

For many people, the solution to living in the polluted East has been a simple matter of moving to the relatively unsullied West. One of the meccas of the modern-day westward movement has been Seattle, Washington, a fog-swept city that lurks between distant, snowcapped mountains and the rolling waters of Puget Sound. Ironically, Puget Sound—amidst whose dark blue waters you can frequently see pods of killer whales and groups of seals plying the waves—rivals Boston Harbor for the grand prize as the nation's most polluted and dangerous body of water.

Puget Sound

Puget Sound sprawls among the rolling hills and forested islands that have become home to Seattle and Tacoma, Washington, and to Victoria, British Columbia. Its deep blue surface is punctuated with humpbacked islands that, observed from a distance, seem part of a pristine world. But Puget Sound is not pristine. Indeed, some of its harbors are among the most polluted bodies of water in the nation.

The sources of pollution here mirror those that have poisoned Boston Harbor. From streets comes storm-water runoff, bringing with it fossil-fuel contaminates from such sources as service stations and street surfaces and adding to them chemicals and fertilizers from car washes, suburban gardens, and other sites. Seventy industries and sewage-treatment plants along the sound are each permitted to discharge more than 100 million gallons of effluent per year. The major industrial sources of pollution include pulp and paper manufacturers, petrochemical facilities, aluminum refineries, and wood-treatment plants. As a result, toxins build up in the harbors that receive the runoff. In one case—the Harbor Island drainage system—the mud bottom is so badly polluted that dredged sediments can be smelted for lead, which constitutes 30 percent of the sediment. So contaminated are some portions of the sound that one of them, Commencement Bay, near Tacoma, was the first underwater site added to the Superfund list, a program created by Congress nearly 20 years ago to designate and clean up the nation's most polluted areas.

Evidence of the grave pollution in Puget Sound has been accumulating for at least 15 years. Much of the evidence concerns observations made on the health of local marine wildlife. It is difficult to pinpoint the source of pollution problems in most fish species, because fish swim from place to place, making it impossible to correlate physical condition with location of catch. But one fish, the English sole, is territorial. Because the English sole stays in one place, living on and under sediments, scientists can correlate evidence of damage from contaminants with toxins found in those sediments where the fish lives. English sole from areas of Puget Sound with highly contaminated sediments frequently have liver cancer. Studies have shown that English sole and other flatfish from various parts of the sound also experience reproductive failures, apparently because of toxic contamination.

When cancerous lesions were first found in the livers of English sole, people refused to believe that pollutants were the cause. They tried to deny that the problem existed. "When we discovered these problems years ago, about 15 or more years ago, people said it wasn't so," says biochemist Donald Malins, who has studied English sole. "People would say, 'You scientists, I mean you'd better go back and take a look at your notebooks, you're crazy. This is a pristine environment. With these tumors, I mean, what is it? It's nothing but alarmism. You're just an alarmist.' Well, it turned out that they reflected serious pollution problems in Puget Sound. We've got Superfund sites in Commencement Bay and in Eagle Harbor. It is not a fantasy of mine. It's not an issue of being an alarmist. It's a real problem."

The reality of that problem is becoming more and more apparent. Commercial harvest of shellfish has been restricted in more than 16,000 acres of shellfish beds since 1981. PCBs and pesticides have been found in killer whales at levels high enough to pose a mortal threat to the animals. High incidences of premature birth and the death of seal pups in some areas around the sound may be related to toxic contamination. At several seabird nesting sites, researchers have reported pesticide-induced eggshell thinning and reproductive failure in glaucous-winged gulls, pigeon guillemots, and great blue herons. Some of the highest levels of DDT contamination ever reported worldwide have been found in these three species in Puget Sound.

Pollutants reach these animals through the food chain. When toxins enter the sound, they either remain in the water or attach to sediment particles. How big a load of toxins the sediment can carry depends on the type of sediment. Coarse sediments, such as sand, have a smaller amount of surface area per gram than do fine sediments, such as mud. A gram of mud has a surface area of 20 to 30 square meters, compared to 0.1 square meter per gram of sand. The muddy bottom of Puget Sound, enhanced by the silt brought in from surrounding rivers, is ideal for carrying large amounts of toxic pollutants, such as copper, mercury, silver, and other metals and chemicals that bind to the surface of mud particles.

The mud particles are eaten by various minuscule animals that live in the mud of the sound's floor. These animals eat 50 to 5,000 particles of mud per second, consuming up to 1,000 times their body weight daily. In the process, they may strip off the toxic contaminants and digest them. The toxins become part of the animals' tissues, to be consumed by larger animals, which are eaten in turn by still larger animals, until those original toxins finally settle in the tissues of birds, seals, and whales. And, of course, humans. We cannot hope to escape

Water Quality Specialist Arlen Walker runs tests for Seattle's Metro Environmental Laboratory. Samples from local industries are checked to ensure compliance with wastewater dumping laws.

the effects of the toxins we put into the water. "In other words, it comes back to haunt you," says Malins. "It's coming back to you because human beings are part of the aquatic food chain. They eat fish."

Fortunately for area residents, there is reason for optimism about the future of Puget Sound. Virtually no part of the sound is free from contaminants, but many areas are relatively uncontaminated. In almost all areas, pollution of the water itself has been reduced through tougher restrictions on industrial discharges and improvements in water treatment. Recently, overflows from combined sewers (those that carry both wastewater and street runoff) have dropped from an average of 20 billion gallons yearly in mid-century to 2 billion gallons in the late 1980s, with a projected reduction of another 75 percent by 1995. The worst contamination now lies in the sediments that make up the sound's floor, and most of that pollution tends to be centered in localized hotspots, generally in bays associated with cities. There, decades of discharge have caused the buildup of intense concentrations of toxins. It is in these areas that fish and shellfish show the most dangerous effects of pollution.

Unlike Boston Harbor, Puget Sound has been the subject of committed efforts on the part of state and local officials. The Washington Department of Ecology has established standards used for identifying sediments that damage biological resources or pose a threat to human health. The state also has developed a program for cleaning up desig-

nated sites and for safely removing contaminated sediments. Seattle has organized a force of "sewer-ranger patrols" that constantly inspect underground sewer pipes for contaminants in an effort to trace the toxins back to their sources and to stop them before they enter Puget Sound. And instead of dumping treated sewage sludge into the sound, Seattle trucks the sludge into nearby forests and uses it to fertilize trees.

Local citizens also are helping out. Some communities have "adopted" streams, cleaning out debris, planting trees, and building fish ladders so that salmon can pass around dams and other obstacles to reach traditional spawning grounds. Says local activist Debbera Stecher, who with other local residents is helping clean up a stream, "We're cleaning our stream because we have to care. We have to care in our own backyards. Conservation starts here. It starts today. It starts in my own stream in my own park. I had a man ask me, 'Does it really hurt to have a junked car in a stream?' And I asked him, 'Do you really want my shoe in your oatmeal?' Well, we have to keep cleaning the stream up. We can't just say we'll clean it up today and it's going to be clean forever. No. So we start with the kids. We train them from the word *go* that this is what we need in order to keep the quality of life we want. We have to keep our streams clean. We have to keep our waterways clean. And once we do that, then it becomes a normal part of everybody's life."

Arcata's man-made marshes, which are adjoined by parklands and protected areas for wildlife, have drawn hundreds of bird species and mammals. The city even celebrates its *au naturel* sewage-treatment system with a "Flush with Pride" festival.

Local citizens must not only *solve* the problems caused by pollution but also *prevent* them. Pollution, after all, begins with each of us, with our own individual actions. "We have been tossing chemicals, all of us—industry, individuals—into Puget Sound either by runoff from the land or transfer from the atmosphere or simply by dumping things over the sides of boats and so on," says biochemist Donald Malins. One outstanding experiment in which city and citizen have learned to cooperate to solve pollution problems is under way a few hundred miles south of Seattle, in a California town called Arcata.

A Wetlands Wastewater Treatment Center

Arcata, California, a timber and fishing town on Humbolt Bay, some 250 miles north of San Francisco, is home to about 15,000 souls. Many of those souls moved to Arcata to escape urban life and run-of-the-mill late-twentieth-century life-styles. In Arcata, those who wish to do so can return to fabled days of yesteryear, provided that the year is roughly 1965. Men wearing long ponytails fit in comfortably with the social milieu. The many who cruise through town in Volkswagen minibuses conform perfectly with local traffic aesthetics. But if some of the townfolk seem stuck in the 1960s, the town itself has shown real innovation.

It began in 1974, when state bureaucrats told Arcata officials that their town and two nearby cities were going to host—and help pay for—a massive sewage-treatment plant that would clean up the wastewater the towns were dumping into Humboldt Bay. The plant would cost about $56 million. The townspeople were not amused. Boatmen feared that they would snag anchors on pipes laid across the bay. Farmers envisioned fields crisscrossed with sewer pipes. Almost everyone thought about the doubling in sewage rates that would soon descend on them.

The project received a stay of execution when a citizens committee in the town of Manila sued to stop the project and brought it to a halt for two years. Folks in Arcata used that time to come up with an alternative plan. City officials tossed around a variety of ideas until finally they came to one proposed by Humboldt State University professors George Allen and Roger Gearheart in conjunction with Frank Klopp, Arcata's public works director. Their plan called for building a wetland on a coastal site that, at the time, was occupied by abandoned, dilapidated lumber mills. After initial processing, wastewater would be channeled through the wetland, where plants and bacteria would filter it naturally before it flowed into the bay. The city decided to adopt the plan, but ran into state opposition. Allen, Gearheart, and a city councilman spent two years going to regional meetings of the state Water Quality Control Board before finally winning approval for a pilot project. They completed their wetland wastewater treatment center for $3 million less than Arcata's portion of the big project would have cost, and it has been functioning well ever since.

Arcata's treatment center is like that of other cities only in the way it begins: Wastewater enters sediment tanks in which solids are removed and disinfected. The sludge is then removed and processed for use as fertilizer. The liquid waste is pumped into oxidation ponds where bacteria break down organic materials. After a month of that, the water flows into the artificial marshes, where plants such as cattails and bullrushes and the bacteria that grow on their stems further purify the wastewater. Some of the water from the marshes is piped into the city's salmon hatchery and mixed with seawater as part of an experiment to create a breeding colony of salmon. It is known that salmon fry learn to identify the place where they were born by its water chemistry. The young salmon then go to sea and, when they reach breeding age, follow the chemical track back to their spawning ground. Arcata hopes to establish its own salmon fishery and so turn the sewage-treatment marsh into a moneymaker.

Most of the treated water is returned to the primary treatment plant, where it is chlorinated as required by state law and then dechlorinated before discharge into the bay. City officials have stated that the water is clean enough to discharge directly out of the marsh but that chlorination is done merely because the law requires it.

The 154 acres of marshes and lagoons that constitute Arcata's waste-treatment plant do more than treat water. They also provide homes for otters, falcons, gulls, owls, waterfowl, and ospreys—in all, some 200 species of birds have been tallied there. The treatment center is also a city park and has become Arcata's claim to fame because it is a big tourist draw. The town even sponsors an annual festival in honor of the wetland. The festival is formally called Waterfront Days, but it is perhaps better known as the "Flush with Pride" festival. Residents show their support by wearing T-shirts that feature the festival's logo—a salmon leaping out of a toilet upon which perches a great blue heron.

The treatment center/park has drawn interest from other towns. Says Public Works Director Klopp, any town with basic sediment tanks and room for some wetlands can duplicate Arcata's project and save money in the long run. It is an elegant, imaginative approach to an old problem and carries promise for better wastewater treatment in municipalities all over the nation.

Strange New World

Who, a few generations ago, could have reasonably believed that toxins generated by industry and domestic life would make the fish of Boston Harbor too poisonous to eat? Who could have guessed that nearly half the nation's shellfish beds would be closed in the name of human health? Who might have anticipated that virtually every estuary along Texas' Gulf Coast, its shores encrusted with oil refineries and other heavy industries, would one day be, for all practical purposes, biologically moribund? Who might ever have thought, even a quarter century ago, that barge canals dug by the petroleum industry would dice up the Louisiana Delta so badly that the state would lose 50 square miles of coast *yearly,* the sediments washing out to cloud Gulf waters? Who would have guessed that the Mississippi River would one day carry such vast quantities of agricultural fertilizers and

Eerie lights paint the night at an oil refinery outside Houston, Texas. Oil refineries and petrochemical plants line the corridor from Houston to Galveston. "Oil Alley" supplies the lifeblood of the local economy but has sounded the death knell for numerous animals.

industrial and domestic toxins that fishermen would need to be wary of eating the fish they caught in it, that the remaining delta lands of Louisiana would be poisoned by contaminants flowing from the north? Or that Chesapeake Bay's multi-million-dollar fisheries would be threatened by collapse as agricultural and urban pollutants contributed to population declines and toxicity in striped bass, shellfish, and crabs? What does it tell us that something so vast as the ocean has become so polluted that Atlantic coastal bottlenose dolphins have become the most toxic mammals on record?

There are no answers to these questions, except perhaps the feelings of dismay, sorrow, and anger that arise when contemplating how this has come about. But, ultimately, those feelings must be directed not only at lethargic governments, inert politicians, and reckless industries, but also at ourselves. Each of us contributes to the decline of the coastal environment every time we pour paints, oils, photochemicals, food wastes, and myriad other products down our drains, every time we treat our lawns and gardens with fertilizers and pesticides, every time we turn on our car engines and send fossil-fuel effluents into the air, to be gathered up by falling rain and deposited in lakes, rivers, and seas. If there are solutions to the problem of coastal and oceanic pollution, they must come from us all. They will lie in changing our daily habits and in insisting that government and industry wake up to the challenge posed by a sickening environment.

Fortunately, people are rising to meet the problem. In many ways we lie on a cusp in the development of human society. Better than any people before us, we recognize the effects that our actions are having on our global habitat. We also are beginning to understand all that we have to do to correct the damage caused by our forebears. The challenge today is translating that recognition and that understanding into suitable action.

It is encouraging, of course, to know that we have begun to take action. We can see it, just as we can see environmental problems, all over the nation. We see it in Boston Harbor, where federal law is forcing city officials to address the dangers that lie in its offshore waters. We see it in the citizens of Washington State who have taken it upon themselves to clean up streams. We see it in Terry Backer, who has made personal sacrifices for a cause that must once have seemed alien to him. We see it in Arcata, California.

We see it, on a grand scale, in the New York Bight, a triangle of ocean bordered on the north by Long Island and on the west by the New Jersey coast and extending roughly a hundred miles out to sea. This is a vastly polluted area, a place long used for dumping thousands of tons of urban wastes daily. It is also the home of the world's largest garbage dump, the Fresh Kills landfill on Staten Island, a borough of New York City. Opened in 1948, the landfill operates six days a week and takes in about 17,000 tons of residential waste daily. It covers 3,000 acres and leaks thousands of gallons of toxins into surrounding waters every day. Solid wastes falling from garbage barges and trucks and blown by the wind end up in surrounding waters, fouling New Jersey shores and forcing beach closings. But efforts are under way to modify the Fresh Kills to prevent the escape of solid wastes and to stop the seepage of toxins, and plans are afoot to close Fresh Kills in the near future. The New York Bight is also being monitored by a combination of federal and state agencies that have developed a plan to improve control of contaminants there. For the first time in history, concerted efforts are being made to correct New York and New Jersey's mutual coastal problems at an official level.

Similarly, cleanup is under way in Chesapeake Bay. Changes in land use in Pennsylvania have helped clean up rivers that bring toxins to the bay, and both Maryland and Virginia have enacted laws that reduce the loss of the wetlands that protect river and bay quality. The Potomac River, shared by Maryland, Virginia, and the District of Columbia, has benefited from new sewage-treatment facilities that have reduced pollutants such as phosphorus from more than 8 million pounds yearly in the 1960s to 63,000 pounds in the late 1980s, a 99-percent reduction in the disposal of fertilizers such as those that have killed portions of Long Island Sound. The citizens action group called the Chesapeake Bay Foundation has labored successfully for more than a quarter century to stimulate regional politicians to take action to protect the bay. Special restrictions on the taking of striped bass are helping to end the overkill largely responsible for diminishing bass populations.

Far down the East Coast, Florida's Bruce and Susan Bingham, a married couple who were appalled by the amount of garbage they saw thrown overboard during a coastal cruise, helped start a campaign to

◄

An osprey with the catch of the day prepares to dine on Maryland's Eastern Shore. The birds have made a comeback since DDT was banned, but now they may be faced with more subtle threats.

Bob Schoelkopf feeds a gray seal, one of many visitors at the Marine Mammal Stranding Center he directs in Brigantine, New Jersey. Wildlife is rehabilitated at the center and either released or sent to zoos. Many more animals wash ashore either too sick to recover or already dead.

Trash on a New York City street awaits the long journey to Fresh Kills landfill on Staten Island. Trashologists are finding that landfills have a marvelous capacity for preserving trash for decades. Most major landfills are either near capacity or well over capacity. The job of finding sites for new landfills is not for the faint of heart.

reduce the amount of plastic wastes and other debris dumped into their offshore waters. Recalling their cruise, Bruce says, "It was the perfect night. Full moon and stars and the ship was rolling gently, and the next thing we know, kersplash, kersplash! And there's a wake of garbage as far as we can see. It was actually coming out of both sides of the ship at once." Adds Susan, "Coming out of the side of the ship were boxes and plastic bags just being thrown out in succession, and we looked back behind us and there was a trail of garbage floating behind it. And it was really shocking to see." So shocking that the duo mounted an intensive media and letter-writing campaign that eventually persuaded one cruise line to stop its garbage dumping.

Such actions by citizens and government show a growing commitment to a cleaner world and offer hope and encouragement in the face of the critical environmental hazards and degradations that we see around us. This conflict of hazards and hope was perhaps best summed up by actor Ted Danson, who founded the American Oceans Campaign to promote cleaner oceans. Danson discovered his cause after a visit with his daughters to Santa Monica beach ended abruptly because the beach had been shut down because of pollution. As host of the Audubon Television Special on coastal pollution, Danson said, "I have learned that our coasts are in peril and must be saved. I have seen the beginnings of a national change of attitude, a growing awareness that this problem is hurting all of us and that we all need to pitch in to solve it. I know that we need to pressure our leaders to make the environment a major national priority, but I've learned that government alone cannot solve the problem. I've seen firsthand that individuals *can* make a difference, that they *are* making a difference. The men and women I've met have taught me that we can't be content to forever *clean up* pollution. We must focus on *preventing* it. We need to be careful what we buy, how we use it, and how we dispose of it. These are our coastal waters, and their problems are ours to solve. Ultimately, it's up to all of us to decide what kind of world we live in, and what we leave behind."

A child plays tag with the next wave near the pier at Santa Monica Beach. Much of America's 10,000 miles of beach are fouled by washed-up trash and more insidious forms of pollution. Each year, we add some 16 trillion gallons of effluents to our coastal waters.

SWEETWATER SEAS: AN ILLUSION OF PURITY

Some 500 miles from the Great Lakes lies the confluence of the St. Lawrence and Saguenay rivers. The two rivers are home to a creature that never comes closer than this to the Great Lakes but that nevertheless has been fatefully influenced by the many environmental problems that beset the lakes.

To find this creature, you could try sailing on the ferry that crosses the St. Lawrence between Rivière-du-Loup on the southern bank and St.-Simeon, a little Quebec town nestled in gently rolling hills overlooking the vast sweep of the St. Lawrence a few miles west of the Saguenay. The ferry is a big bathtub of a boat that takes not only passengers but cars as well. It churns steadily across the water with scarcely a bob, comforting to anyone who takes any threat of seasickness personally.

As the ferry passes the Ile aux Lièvres—a long, low island in the dark blue water—you will almost certainly spot something truly remarkable. Fellow passengers will line up along the decks of the ship to get a better look, and they will point and talk among themselves and take pictures.

What they are seeing is a vanishing part of St. Lawrence River history—indeed, a vanishing part of the river's prehistory. They are looking at belugas, called "sea canaries" by whalers in earlier centuries because of the birdlike whistles with which the whales communicate. The belugas appear as white slivers that roll gently out of the dark waters and submerge again with scarcely a ripple. They keep their distance from the ferry, but their brilliant whiteness against the deep blue water—stark white, dazzlingly white, like fragments of the moon—makes them impossible to miss.

They are small creatures, as whales go. They grow at most to only about 15 feet long and never weigh more than a ton. They stay so far from the ferry that they remain little more than chips of ivory, but in the imagination they evoke far greater images.

They recall a time, some 15,000 years ago, when polar bears and walruses lived as far south as the U.S.–Canada border in a chill world dominated by glaciers two miles high. They recall a time, four and half centuries ago, when French explorer Jacques Cartier became the first European to sail up the St. Lawrence. Recording the event, he wrote of

> a kind of fish of which no man has seen or heard . . . they resemble as far as the body and head, a hound, as white as snow without any spots. There are very great multitudes in said river and they live between the sea and the fresh water. The natives of the country call them *adhothuys*.

The *adhothuys* roamed the St. Lawrence more widely then and in greater numbers than they do today. As late as the mid-nineteenth century some 5,000—and perhaps as many as 15,000—lived year-round in the river, descendants of belugas that inhabited a now long-

◄
Sunset over the St. Lawrence River, the final resting place of many Great Lakes toxins.

Beluga whales swim in the St. Lawrence River. The name *beluga* comes from the Russian word for "white." The little white whales were hunted for oil and hides until the late 1970s. Now endangered, they are still hunted by an army of chemicals that inhibits the creatures' ability to reproduce and to resist disease.

vanished inland sea that covered the region traversed today by the St. Lawrence. When the glaciers retreated, the sea disappeared and polar bears and walruses moved north, but the belugas stayed, living in the river, a freshwater population separated forever from belugas that still swim arctic waters. The arctic population has fared well, if numbers are any sign of how a species fares. Though hunted by native peoples, arctic belugas number in the thousands. The St. Lawrence *adhothuys*, meanwhile, have been dwindling away.

What may be the end for the belugas began, as it did for so many other species, soon after the first Europeans arrived. Early whalers killed belugas for oil and hides. The killing accelerated in the twentieth century. Hunters herded belugas into corrals built along river banks where ebbing tidal currents trapped the animals. As many as 100 could be killed at a single low tide. The beluga's numbers began falling, reaching about 1,200 by 1960. They were slaughtered not only for trade but because fishermen thought they were pests, that they competed for fish such as cod and salmon. It has been rumored that the Canadian government even permitted the use of airplanes to bomb the whales, though no official documents have ever corroborated this.

The killing slowed in the 1950s as it became clear from studies on beluga eating habits that the animals did not feed on cod and salmon. During the 1960s and 1970s, an average of about 10 to 20 belugas were killed yearly, and even this stopped when hunting was outlawed in 1979. Nevertheless, four years later the St. Lawrence belugas were added to the Canadian endangered species list. Despite protection, the population has shrunk during the past 30 years to fewer than 500 animals.

Why the decline has continued did not become clear until the 1980s, when biologist Pierre Béland, an independent consultant with the St. Lawrence National Institute of Ecotoxicology, and former government veterinarian Daniel Mardineau started to examine the tissues of belugas that washed ashore dead. What they found was startling. The whales were riddled with pollutants, including carcinogenic PCBs, DDT, and BaP (benzo(a)pyrene, a by-product of such industrial procedures as aluminum refining and the combustion of fossil fuels).

Béland and Mardineau have examined and tested an average of about 15 dead whales yearly since 1982. The levels of toxics found in the belugas are among the highest ever recorded in a mammal, surpassed only by levels found among coastal bottlenose dolphins off the U.S. East Coast. The researchers believe that the pollutants are suppressing the belugas' immune systems, making them susceptible to disease, and reducing their reproductive rate. Necropsies of these belugas have also revealed a high incidence of digestive problems, such as ulceration of the stomach, including perforated gastric ulcers, which had never been seen in marine mammals. The animals also suffer from intestinal ulcers and blockage of the intestine. The rate of tumor incidence is 10 times that found in dead whales examined elsewhere. Some of the tumors are cancers never before found in whales or their relatives, except in harbor porpoises. Hundreds of dead whales have been examined without the observation of a single cancerous tumor, but among river belugas, 5 out of 40 carcasses have cancer.

Few of the females examined—barely a third—appear to be in reproductive condition. In contrast, about two-thirds of dead beluga females examined in Alaska showed signs of having reproduced. Many females from the St. Lawrence have lesions on their mammary glands, which suggests they could not feed potential young. Even if they could reproduce and nurse, many of their young would still be doomed to an early death. Pollutants such as PCBs are concentrated in fats. About 40 percent of a beluga's weight is blubber, and some of that fat—with its pollutant load—goes into the rich milk that mothers produce. Newborn calves—already contaminated by toxins that reached them in the uterus—receive large doses of PCBs and DDT when they feed on their mothers' milk. During their first year of life, calves may accumulate higher levels of contaminants than do older males, which, because they lack the female's means for getting rid of some toxins in milk, are the segment of the population most likely to build up high pollution loads.

The total effect can perhaps be summarized in these two facts: Belugas in the St. Lawrence appear to be dying at an early adult age at a faster rate than would be expected for a long-lived mammal, and the birth rate for the population is about half what is observed in arctic beluga populations.

Many of the problems found in the belugas are identical to those observed in laboratory animals that are fed contaminated food experimentally. Moreover, bladder lesions found in the whales are like those found in workers exposed to BaP at the Alcan aluminum plant in Chicoutimi, Quebec, a town that lies on the Saguenay River, which is frequented by the whales.

At his office in Rimouski, Quebec, biologist Pierre Béland discusses the fate of the St. Lawrence population of beluga whales. These river-bound cetaceans once numbered in the thousands, but have been reduced to a mere 500. Béland's autopsies of more than a hundred dead whales since 1982 have shown the animals to be riddled with tumors and cancers.

The sources of the pollutants that plagued the whales are many. According to Béland, heavy metals reached the river from pulp and paper plants at Rivière-du-Loup. Large amounts of mercury, fluorines, and other carcinogens were released from one of the largest aluminum-production facilities in the world.

Ominously, some of the potent chemicals affecting the whales are coming not so much from the river as from areas some 500 miles away. One chemical is mirex, a powerful insecticide that was once produced at a plant in New York on Lake Ontario. It was banned in the early 1970s after it leaked into the lake in quantities that killed thousands of fish, but it is still turning up in belugas at levels 1,000 times higher than can be found in sediments near the old production site. How is it getting to the whales?

Béland speculates that it is reaching them through the fish they eat. He is particularly interested in the eels that migrate from Lake Ontario each year on their way to breeding waters in the Sargasso Sea. The whales, he believes, could be ingesting mirex—present in the eels at about a tenth the level observed in belugas—by feeding upon the eels during their two-week migration period. In effect, the eels are living transport systems, bringing chemicals hundreds of miles from their source and into the mouths and bodies of distant beluga whales.

It is not surprising that the fate of animals far down the St. Lawrence should ultimately be linked to events in the Great Lakes. The lakes are the single greatest source of most of the St. Lawrence's flow, and they lie in one of the most heavily developed parts of the continent, an area

A papermill smokestack puffs through the night on the Canada side of the St. Lawrence River. Pulp and paper plants along the river and its tributaries have contributed heavy metals to the load of toxics that flush from the Great Lakes.

bordered by farms and dotted with factories and refineries. Nearly half of the hazardous-waste sites put on the U.S. Environmental Protection Agency's priority cleanup list are found in states bordering the lakes. That the lakes should contribute to the poisoning of distant river waters is merely a single measure of the deadly problems that hide beneath the lakes' vast waters.

A Great Lakes Primer

Geologists believe that the Great Lakes did not exist before glaciers advanced south some half-million years ago. The glaciers were produced by a global chill that allowed snow and ice to accumulate year after year in the far north until finally the immense weight of the heaped ice caused it to squeeze itself south, much as a piece of soft clay spreads out if you press down on it. The moving of the glaciers ever southward must have been one of the greatest natural spectacles the Earth has ever known. The sheer wall would have creaked and groaned as the force of its weight pushed it across northern reaches. Huge splinters must often have broken off the front of the glacial wall, fragments the size of city office buildings that would have rocked the Earth with the shock of their falling. Ahead of the glaciers was all the rubble that the vast ice flow pushed aside with impunity, scraping off trees and shrubs and boulders and rock and the surface of the Earth itself, rolling it all up like a carpet.

Four times the glaciers have eased out of the north, and for all we know they may come again. The last glaciation faded only 10,000 years ago, and the shortest period between glaciations is five times that length. Somewhere down the line, in generations too distant from ours to contemplate, our descendants may witness the rebirth of an Ice Age. Meanwhile, the glaciers have etched into the face of the Earth monuments that have outlived our memory of the ice. One of the greatest is the Great Lakes.

The Great Lakes are so huge that early explorers thought they were freshwater seas. Lake Superior alone is astoundingly big for a lake. Holding half the water in the Great Lakes system, it is the world's largest body of fresh water, containing more than 10 percent of the world's surface fresh water. It covers an area that could nearly encompass Connecticut, Massachusetts, Vermont, and New Hampshire—roughly 32,000 square miles—and its deepest waters measure more than 1,300 feet to the bottom. Even the smallest of the lakes, Ontario, is 50 miles wide and nearly 200 long, with a surface area larger than that of Delaware and Connecticut combined. With the other three lakes—Erie, Huron, and Michigan—the five hold 95 percent of the surface fresh water in this nation—6 quadrillion gallons—enough to cover the entire lower 48 states 10 feet deep.

GREAT LAKES POLLUTION

THE ISSUE The Great Lakes are heavily polluted with many toxic industrial and agricultural chemicals as well as urban wastes and runoff.

THE CAUSE Cities have dumped wastes, often untreated, into the Great Lakes since the first days of settlement. The many industries that sprang up around the lakes have filled them with toxins, such as PCBs and heavy metals, that sink into the lake bottoms. Though the water is clear, the toxins persist in the sediments, where they enter the food chain. Airborne toxins are another problem. Pesticides have reached the Great Lakes from hundreds of miles away by riding prevailing winds.

EFFECTS ON WILDLIFE Many bird species, including cormorants, gulls, and pipers, are showing great numbers of birth defects, such as fatally crossed bills and loss of appendages. The beluga whales of the St. Lawrence are so heavily contaminated with pollutants that their reproductive rate is collapsing. Many shellfish beds are heavily contaminated, as are fish. Some fish species are failing to reproduce. Data suggest declines in the number of aquatic mammals, such as otters, and in bald eagles that live along lake shores.

EFFECTS ON HUMANS Some data indicate increased instances of miscarriage in women who regularly eat Great Lakes fish. One study suggests that children born of women who eat the fish show slower intellectual development than children of women who do not eat Great Lakes fish. At this stage it is difficult to isolate specific effects on people. It is known that eating a single lake trout from some of the lakes can give you a dose of toxic chemicals equivalent to drinking lake water for roughly two full lifetimes. Contamination of fish and shellfish has led to declines and collapses in commercial fishing and in the tourist industries.

EFFECTS ON HABITAT Sediments become contaminated, allowing toxins to enter the food chain at the bottom and work their way to the top, from microbes to eagles and humans. Coastal regions, because of the toxicity of fish, are becoming hazardous to eagles and aquatic mammals that live there. Toxins that enter the lakes eventually travel down the St. Lawrence, poisoning the entire ecosystem and possibly contributing to ocean pollution where the sea meets the Gulf of St. Lawrence.

The Great Lakes were first discovered by wandering hunter-gatherers who came out of Asia at least 15,000 years ago, moving into the New World over a land bridge that connected Alaska to Siberia. The land bridge was itself a by-product of the glaciers, created when the seas were lowered because so much of their water was tied up in ice. These first explorers left no record of their exploits, save the oral leg-

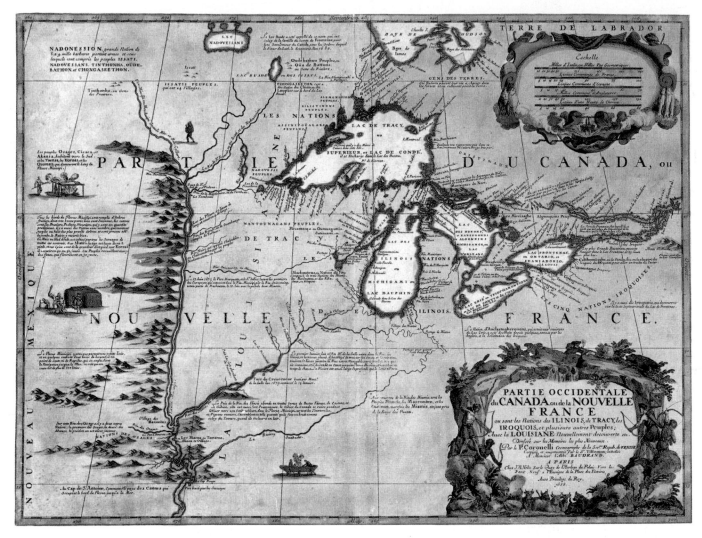

This 1688 map of the Great Lakes by the Italian Coronelli was widely circulated in France and Italy.

ends found among such peoples as the Winnebagos. When the Winnebagos, who lived in the sixteenth and seventeenth centuries on the shores of Lake Michigan, encountered the first European expeditions, they told the newcomers that their people had lived long ago on a big lake far to the west. This was probably Lake Agassiz, which several thousand years ago covered large portions of Manitoba, Ontario, Minnesota, and North Dakota. It silted in long ago, but the Winnebago culture carried its memory down through history.

The first record of European arrival in the Great Lakes area dates from 1651, when Samuel de Champlain nosed a boat through the North Bay gap and found Lake Huron at the other end. He dubbed it *La Mer Douce*, the Sweet Sea, because its water was fresh. Then he sailed among its many islands as he searched for the passage that would take him across the Northwest to Cathay. He was disappointed on all counts, but he had added the first of the lakes to the historical record.

Ernest Hemingway once wrote, "A continent ages quickly once we come to it." Nothing truer can be said of the Great Lakes. It took roughly half a century after Champlain's arrival at Lake Huron for the rest of the lakes to be met by Europeans, but by that time the process of aging was speeding up.

Perhaps its earliest agents were two teenagers named Pierre Esprit Radisson and Médard Chouant, Sieur des Groseilliers. Youthful and

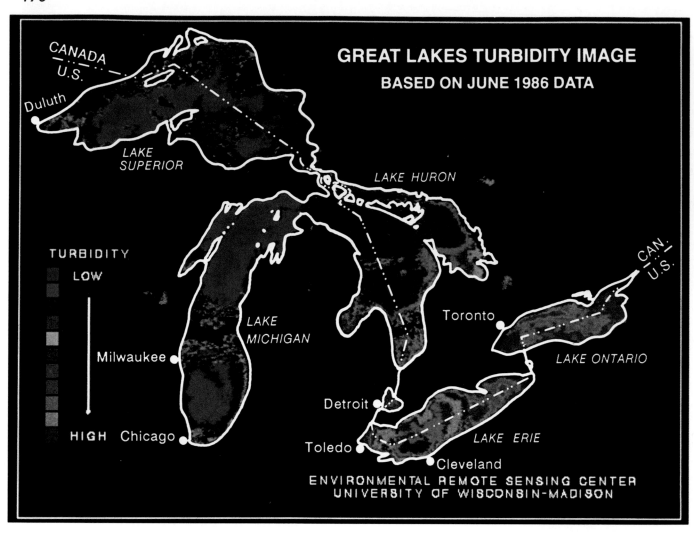

GREAT LAKES TURBIDITY IMAGE
BASED ON JUNE 1986 DATA

TURBIDITY
LOW

HIGH

ENVIRONMENTAL REMOTE SENSING CENTER
UNIVERSITY OF WISCONSIN-MADISON

apparently ambitious, they sneaked out of Quebec and went in search of the Northwest Passage. As did many before them and after, they failed in that quest. But they did return to Quebec with 50 canoes packed with furs.

Furs! What a great idea! Or so thought the early European explorers, who until then had seen the New World as merely an obstacle to travel to the Far East. But Radisson and Groseilliers' enterprise suggested that profits could be made from the continent.

The Company of New France, the official French agency for New World commercial development, confiscated the teenagers' furs, so the duo went off in a huff and helped the British start the Hudson's Bay Company, which eventually put the Company of New France out of business. The Hudson's Bay Company, with the aid and abetment of generations of trappers, nearly put a few species out of existence, too. The onslaught of the fur trade nearly extinguished North America's beaver, river otter, and sea otter, among others, and did wipe out the sea mink.

Despite the fur trade, the Great Lakes themselves made it through the eighteenth century relatively intact, primarily because hostile Native Americans made the region uncomfortable for settlers. But that obstacle was soon overcome as a result of political pressures. The

The colors on this satellite photograph of the Great Lakes represent water turbidity resulting from suspended sediments, river inputs, phytoplankton growth, and other factors. Land and clouds are black. The clearest water is shown as deep purple to blue; more turbid water as yellow to deep red. The image was obtained on July 21, 1986, from the Advanced Very High Resolution Radiometer on the NOAA-9 polar orbiting weather satellite.

newly created United States received portions of British lands south of the Great Lakes as part of the settlement ending the Revolutionary War. The politicos of the time wanted settlers—good American settlers—to move into the region quickly to solidify the new nation's claim to the area, and with ardent government support the Native Americans were soon vanquished.

Settlement began in the Firelands, so called because they were given to people whose property had been burned in the Revolutionary War. These lands are today the portion of Ohio that borders Lake Erie. Cleveland was settled in 1796, and other cities followed early in the next century—Ashtabula, Sandusky, and Huron were founded by 1810. In 1820, Milwaukee was a nascent community, inhabited by two American families and a village of Native Americans. It was founded at the mouth of what was then called the Milwacky River. By 1833, Milwaukee's American populace had expanded by one entire citizen. But the boom was soon to come. By 1845, Milwaukee had 10,000 residents. The number leaped to 46,000 in 1851.

Chicago followed much the same pattern. In 1830, Chicago had perhaps a dozen families to its name. Before the end of the decade, the town had 5,000 residents. By the time the Civil War had started, the population was more than 100,000. And it was only one of many towns being built on Great Lakes shores, both north and south of the international border. By midcentury the lakes were studded with towns and villages, hamlets and cities.

With the coming of the settlers the aging process naturally accelerated. This was evident in what quickly happened to the rivers that drained into the lakes. These rivers are characteristically short and sluggish, making them miserably poor conduits for city wastes. In Ohio, rivers described as clear by settlers in 1825 were called turbid only a quarter century later. The Chicago River was described in J. B. Mansfield's 1899 book *History of the Great Lakes* as "defiled and putrescent with sewage and filth." When it became clear that the rivers would not suffice as sewers, sewage was simply piped directly into the lakes. Townspeople also dredged canals in the rivers along which cities were built and drained wetlands to provide more land for construction. Both activities added sediment to the rivers.

Settlement changed the face of the land, too. Forests—dominated by hardwoods in some areas, pines in others—pressed against the Great Lakes. The first settlers cut trees simply to be rid of them, to open land for farming. Those bent on building cities cut forests for construction materials. But logging took a serious new turn in the 1830s, when several converging historical developments rapidly doomed the forests.

One development was the opening of the Erie Canal, which made it easier to transport goods to eastern markets. Another was the mechanization of the logging industry through the invention of the steam engine and such implements as the circular saw. And, perhaps most important, settlement was gradually moving west into the prairies, creating a greater demand for wood for construction.

The forests of Michigan, around Saginaw Bay, were the first target of loggers who had already exhausted forests farther east. The first lumber mill on the bay opened in 1835. Sixty years later, the area

A leaf floats atop lilies and duckweed at Cooper Conservation Area outside Cornwall, Ontario. Ninety-five percent of U.S. surface fresh water lies in the Great Lakes, where superficial improvements in water quality have masked a persistent toxic threat.

boasted 100 mills. Similar development occurred throughout the region. As forests fell, mills sprouted at towns such as Augusta and Alpena, Menominee and Marquette. Millions of board-feet were cut yearly as loggers swept into Wisconsin and Minnesota. The cutting was rapacious and wasteful. Some trees were cut just for their bark, which was used in tanning hides. Trees were cut illegally on land held by absentee owners. Veterans whose benefits included cheap land in the lakes region sold their homesteads to logging companies for a quick buck. Vast areas were denuded, and some remain barren to this day. The Kingston Plains in Michigan's Upper Penninsula are an example. The tall, straight pines that enshrouded the plains were wiped out at the turn of the century. Today, only their gray stumps remain, stark tombstones rising from weeds and shrubs. The forest, for unknown reasons, has never grown back, perhaps never will.

By the 1920s the logging boom was over. The forests were gone, lost to an era of exploitation in which no one—or at least no one among those in charge—could see that the forests were finite. Everyone thought there would always be one more forest to cut and that cut forests would grow back. One effect of the logging was the choking of streams with sediment that eventually reached the lakes themselves. Tons of sawdust and other logging by-products were also dumped into streams. The sawdust did two things. It settled to the bottom of streams and suffocated the plants and animals that lived there, and it used up oxygen as it decayed, just as algal blooms use up oxygen as algae sink to the bottom and decay. This destroyed the food webs of many rivers and streams, contributing to declines in many fish species, including chub, whitefish, blue pike, and the Atlantic salmon of Lake Ontario.

Industry came to the Great Lakes early in the twentieth century, leading to further degradations. The refining of steel and aluminum and the manufacture of pesticides at many sites along lake shores led to an influx of deadly toxins. Some 300 toxins have been identified in lake waters, including such hazardous materials as lead, nickel, chromium, cyanide, fluorides, and arsenic. The steel industry dumped so much asbestos into Lake Superior—the most pristine of the five lakes—that the floor of the lake's entire western arm is buried under the carcinogenic substance. The shipping industry has contributed to pollution with leaked oil and to erosion of rivers and shores as the wakes of passing ships have pounded the banks. About a third of Great Lakes shores suffer from erosion, primarily in areas near shipping lanes.

Lake Erie has suffered the most damage from contact with urban populations. Lake Erie is relatively shallow—at its deepest it is only 210 feet deep, and it averages less than 60 feet—and lies on fertile soil full of nutrients. These two conditions make it a perfect victim for death by nutrient overload. Such a death came soon after development intensified. By the middle of this century Erie was being bombarded with nutrients that cause algal blooms. The nutrients came from a variety of sources, including agricultural runoff and such substances as detergents in city wastewater. Every major river feeding into Lake Erie was grossly polluted. Two cities at opposite ends of the lake—Detroit, Michigan, and Buffalo, New York—had dumped virtu-

A view of the Buffalo River shows the industry and shipping that claimed the waters in the early twentieth century. In those days the Great Lakes were a vast dumping receptacle that was thought to carry industrial refuse to some place where it magically disappeared. Decades later, we are paying dearly for this fantasyland mentality.

ally untreated sewage into the lake for decades. Every day the Detroit River carried 20 million pounds of pollutants into Lake Erie, including nutrients such as phosphates and nitrogen and toxics such as phenols, ammonia, and chlorides. But the Buffalo River made the Detroit look unsullied by comparison. It was so stifled by pollutants that even relatively toxin-resistant species, such as sludge worms, could not live in its sediments, while its waters were packed with coliform bacteria, creatures found in human feces. The Cuyahoga River, where Cleveland lurks, was slick with oils and other chemicals so flammable that it was declared a fire hazard and would in fact achieve national infamy when its Cleveland section burst into flame in 1969.

In the late 1950s it became clear that Lake Erie was ailing. One 2,600-square-mile portion of the lake was found to have no dissolved oxygen in its bottom waters. By the 1960s, lake shores were heaped with detergent suds, rotting algae, and dead fish.

But Lake Erie was not an exception. Lake Huron's Saginaw Bay was also ailing, showing all the symptoms of Lake Erie. So was Green Bay, in Lake Michigan.

The first rumblings of a public outcry arose in the late 1960s. Eventually, political activists pushed through legislation to begin protecting the lakes. In 1972, enactment of the Clean Water Act set guidelines for water cleanliness and cleanup and provided federal aid for wastewater-treatment improvements. Canada, in the same year, enacted its Water Act, which set standards for the amount of phosphates permitted in sewage effluents. A cooperative agreement for Great Lakes cleanup, the International Great Lakes Water Quality Agreement, was also signed by the United States and Canada in 1972 to help control nutrient phosphate flow into the lakes. In addition to these potentially sweeping developments, the lakes states, provinces, municipalities, and industries reached a series of agreements for improved cleanup and lake protection.

The new developments helped bring positive changes to the lakes and their tributaries. By the mid 1970s, fish swam once again in the Detroit River. The amount of phosphate in waters flowing into Lake Erie was halved. Algal blooms disappeared, and the lake went from green to blue. Even the Cuyahoga changed color—from black to light brown, not exactly perfect but an improvement. If nothing else, the river was no longer a fire hazard.

Lake Erie became a symbol of hope for all the Great Lakes, a sign of what society could do if only it tried. Unfortunately one highly important fact had been overlooked—what you don't see can, nevertheless, kill you. Despite the increased clarity of the waters, and the return of fish to lakes and rivers where sedimentation and pollution had killed them off, all was not well. In fact, much was still quite sick. An invisible hazard lurked in the vast waters of the Great Lakes, and it is still there today, ominous, waiting.

◀

The Seaway Queen docks at Toronto on Lake Ontario, Canada. Through the years shipping has added oil, chemicals, and rust and paint particles to the Great Lakes' toxic brew and has increased erosion of river banks and shores.

Hazardous Shores

Some 40 million people live in the Great Lakes region, accounting for 13 percent of the U.S. population and about 30 percent of Canada's.

Since the two nations signed the 1972 Great Lakes Water Quality Agreement, they have spent more than $10 billion on sewage plants, yet the lakes' human populations remain exposed to one of the most intensely polluted, dangerously toxic places in North America.

The International Joint Commission was established by the 1909 U.S.–Canadian Boundary Waters Treaty to help settle disputes over boundary waters shared by the United States and Canada and still serves as an advisory body to the two national governments in matters concerning Great Lakes protection (for example, the commission's 1970 report on pollution in lakes Erie and Ontario led directly to the 1972 Water Quality Agreement). Recently, the commission has reported that nearly 300 chemical compounds haunt the lakes and their sediments. Many are threats to humans and wildlife, including dioxins, pesticides such as DDT and mirex, and the seemingly ubiquitous PCBs. These are invisible, odorless pollutants that imbue lake waters and sediments, that are harbored in the tissues of fish and other aquatic species, and that have worked their way through the food web and into the tissues of predatory creatures such as eagles, falcons, otters, and humans.

That these toxic chemicals are still sloshing about in the Great Lakes, in some cases decades after use of the toxins was banned, is ominous, to say the least. These chemicals lend veracity to the old saw that future generations will suffer for the sins of their fathers. Such persistent toxins as mirex, a pesticide capable of killing for years and even decades, are likely to plague human society for many generations to come.

Frogs hunker down in a swamp at Cooper Conservation Area outside Cornwall, Ontario. The lakes are surrounded by pristine-looking areas, but when scientists study eagles, otters, and other animals at the top of the food chain in these border zones, they find low reproductive rates and a high incidence of birth defects.

So what are some of these chemicals? Where did they first come from? And why are they here today?

Perhaps the most frequently mentioned pollutant is the group of chemicals known generically as polychlorinated biphenyls, or PCBs. They first turned up in the 1880s, when German chemists were puttering around with coal tar, a black goo left over from the process of extracting gas from coal. Coal tar offered promise as a source of dyes. Perhaps unexpectedly, they also became the source of compounds called *araclors*. Araclors were dismissed as useless for about 40 years following their discovery, thus giving the world a reprieve from the day when it would be faced with PCBs. The odorless, colorless araclors were thick fluids that were chemically inert and nearly impossible to break down by any of the normal methods, such as the application of heat or pressure. There were about 100 different araclor compounds, and the Germans were not particularly entranced with any of them. The compounds were simply too stable to work with.

Enter a touch of American ingenuity in 1929. American chemists observed that some araclors were perfect lubricants for machinery requiring oils that could withstand heat and pressure. Other araclors were great insulators of electrical transformers or served as ingredients in ink, paint, and plastics. Some could even be sprayed on roads to hold down dust. By then, the American chemists had started calling araclors by a name that reflected their molecular structure, which included a number of chlorine molecules and a couple of benzine rings that formed a biphenyl compound. Thus, araclors became polychlorinated biphenyls, or PCBs. Today, hundreds of different types have been created, some of them potent carcinogens.

PCBs are related chemically to a variety of other famous toxins, such as DDT, which also features a number of chlorine atoms fastened to some carbon and hydrogen atoms. A whole array of polychlorinated hydrocarbon pesticides came into use during and after World War II. DDT perhaps made the biggest name for itself, since the harmful effects of DDT on birds was one of the first signs that the chemical revolution of the late 1940s and the 1950s might have some unexpectedly grim side effects. DDT persisted in the food web, becoming increasingly concentrated at the higher levels. Fish would eat insects or microscopic aquatic organisms that contained small amounts of DDT. The DDT would accumulate and intensify in the fish's tissues as they ingested more and more DDT. Animals that preyed on fish, including humans, would be dosed with the DDT in the fish tissues. The DDT would be locked up in the predator's fat and would build up increasingly over the years in a process called *bioaccumulation*. ·

Mirex, another polychlorinated hydrocarbon pesticide, proved particularly threatening. It was used in the Southeast to kill fire ants, and though never applied in the Great Lakes region, it was manufactured there in Buffalo, New York, and the surrounding vicinity. By the 1970s, mirex proved so deadly that the 1978 update of the Great Lakes Water Quality Agreement called for reducing mirex levels in the lakes to zero, the first time so stringent a limit was ever initiated. By then, hundreds of pounds of the stuff had reached Lake Ontario and rivers near Buffalo. One plant alone lost a ton of mirex. Like DDT, mirex bioaccumulates. By the late 1970s, it was present in the flesh of

A windsurfer cuts the water on Lake Erie near Buffalo, New York, with grain silos in the background. The lakes were once so polluted with human wastes that not even sludge worms could survive. Today, the waters are "cleaner," meaning that recreational swimming is no longer dangerous, but accumulated toxins stored in many fish species pose a health hazard for fresh-fish fans.

◄

lake fish at concentrations several hundred times greater than that thought safe for human consumption.

Another group of persistent chemicals are the PAHs, short for polycyclic aromatic hydrocarbons, by-products of the burning of fossil fuels. One of the better studied of this group is BaP, or benzo(a)pyrene. When eaten, BaP breaks down into chemicals that can attach to DNA, causing genetic mutations that result in cancer. BaPs are especially prevalent in areas near aluminum refining plants. They have been found in high levels in the brain tissues of the St. Lawrence beluga whales.

Chemicals such as BaPs and PCBs and toxic heavy metals such as lead and mercury have been spewed into the Great Lakes for years. Even banned chemicals, such as DDT—outlawed in the United States in 1972—are ever present in the lakes. But even if all toxins were somehow locked out of the Great Lakes area, many of these environmental poisons would continue to accumulate there.

The major reservoir for toxins in the Great Lakes, according to a report prepared by the International Joint Commission, is lake-bottom sediments. Many of the hundreds of lake pollutants are heavier than water and so tend to settle to the bottom, where they mix with mud. Lake sediments even provide a geological history of toxic pollution. Layers of silt laid down in the 1940s and 1950s mark the first appearance of polychlorinated hydrocarbon pesticides, revealing exactly when use of the chemicals first came into vogue. Great quantities of pollutants—particularly heavy metals and PCBs—are laced into bottom sediments. Many of them are extremely persistent, taking decades to break down into harmless chemicals.

Toxins in the sediments would be relatively benign if they stayed there, but they do not. Instead, they seep constantly into the food web that encompasses both wildlife and humans. They do this in the same way that sediment toxins reach fish and other animals in Puget Sound, Boston Harbor, and other contaminated sites: Tiny organisms that feed on silt gather the toxins in their tissues, are in turn eaten by large animals in whose tissues the toxins reach greater concentrations, and so on until, by the time fish enter the food web, the toxins are accumulating at levels that threaten human health. Eating a single lake trout from, for example, Lake Ontario, will force upon you more toxins than would a lifetime of drinking the lake's water.

Toxins also escape the sedentary life when harbors and canals are dredged, as they must be periodically at places such as Chicago, Buffalo, Detroit, Toronto, Cleveland, Green Bay, and wherever else shipping is a big business. Storms also stir up toxic sediments. Runoff from farms and construction sites adds to natural siltation, a process by which rivers carry sediment into lakes and gradually, through the millennia, fill in lake bottoms until the lakes become marshes and the marshes become dry land. This natural process is not cordial to transportation vessels that depend on the harbors and canals as their byways, so shipping companies undertake massive dredging projects every few years to ensure adequate harbor and canal depths for large ships. Every time a harbor is dredged, sedimentary toxins are stirred up and again mix with the water where, until they settle again, they are ingested by any creatures feeding or drinking in the area.

A cotton boll in New Mexico seems an unlikely suspect in the Great Lakes toxics mystery, but pesticides from southern states and even DDT from Central and South America are showing up in the "sweet-water seas."

The other major source of pollutants once posed something of a mystery of scientists who were trying to determine why chemicals such as DDT continued for several years to increase in the lakes even after the 1972 ban and why, even more strikingly, a pesticide such as toxaphene—used to kill boll weevils in the Deep South—was turning up in the sediments of lakes that had not received any influx of sediments from farmlands, let alone from southern cotton fields.

One example is Siskiwit Lake, which lies on Isle Royale in northern Lake Superior. Superior, particularly in the north, where it is surrounded by forests, is the most pristine of the Great Lakes. Isle Royale is home to the national park that bears its name. It is also home to wolves, moose, and a few other large mammals. People have left the island more or less alone, except for park visitors and backpackers. It is not farmed, and it is not sprayed or fertilized to grow crops. By most definitions, Isle Royale is a wilderness, with all the connotations of purity that the word suggests. Yet there, on Isle Royale, is Siskiwit Lake, whose sediments, scientists discovered, are laced with toxins that include dioxins, furans, PCBs, toxaphene, and DDE—a by-product of DDT. Fish tested there in the mid-1980s had double the PCBs and almost 10 times the DDE of fish from Lake Superior itself.

When scientists realized how the pollutants reached seemingly undisturbed Siskiwit Lake, the answer seemed all too obvious: They were brought in on the winds, deposited from the atmosphere. Re-

An older female bald eagle recuperates from a bullet wound at the McDonald Raptor Center, McGill University in Montreal. Eagles, nearly wiped out by DDT in the 1960s, are recovering their numbers in many areas, but not around the Great Lakes.

searchers with the Environmental Protection Agency determined that polychloride hydrocarbon pesticides started appearing in Siskiwit Lake sediments shortly after World War II, since core samples from mud layers older than that showed no signs of the pesticides. The chemicals had been arriving on the wind for perhaps four decades by the time the study was done. The wind is also the likely source of some 80 percent of the PCBs in the rest of the Great Lakes area. In addition, an estimated 15 tons of PAHs reach the lakes yearly on the wind. The wind thus may well be the largest single current source of Great Lakes pollutants.

In the United States, global rotation generally causes winds to come out of the west, but in the summer, winds come in from the Gulf of Mexico, and in winter from the arctic. Pollutants carried to the Great Lakes probably spring mainly from the Midwest and the South, and from as far away as Mexico and even Central America.

Using sensitive equipment, researchers traced the source of the toxaphene found in the bottom of Siskiwit Lake to farmlands near Greenville, Mississippi. The pesticide evaporated from the croplands and moved north over St. Louis and Lake Michigan before ending up at Isle Royale, more than 1,000 miles from home. Most chemicals, however, cannot be so specifically traced. There are simply too many sources. The entire nation is riddled with factories, industries, agricultural lands, urban centers, and private homes that are billowing with poisonous effluents. Every one of the nation's 137,324,000 automobiles is a miniature factory, pumping out toxins. In New Jersey, some 60,000 factory chimneys have been licensed for legal operation. Another 60,000 may be at work illegally. In Texas, perhaps 1,000 different chemicals reach the atmosphere, bound for who knows where. And all the chemicals released by all those sources enter an immense, invisible cauldron in which a venomous witch's brew is being concocted above us and around us, unseen. In the skies, one chemical meets another and another and new compounds are formed, some harmless, some not. Eventually, the 1,000 may be 10,000, or they may be more. As a result, with every breath we take we bring more toxins into our systems. As a result, hardly a man or woman alive does not carry such chemicals as PCBs in fats and other tissues. As a result, our generation is witness to the first known toxic precipitation, created when rain or snow sweeps chemicals from the air and into rivers and lakes.

In many cases, the toxins showering down upon us are the very ones we banned in the United States 20 years ago but permitted American companies—with an insensitive commitment to free enterprise—to continue selling to Latin American nations. For example, most of the DDT turning up in the Great Lakes was probably first unleashed south of the border on crops in Central or South America, used by farmers who bought it directly or indirectly from American pesticide companies. The vulture has come home to roost, and it is, with an ominous glint in its eye, casting its glance over various Great Lakes creatures, animals such as bald eagles, lake trout, and us.

Creatures in Crisis: The Wildlife Factor

Explorers still roam the regions of the Great Lakes, but they are not seeking a Northwest Passage or the riches of Cathay. They are seeking to discover the dangers that lie in wait for us among the contaminants of the Great Lakes, and in that sense they are exploring uncharted seas. They have been looking closely at the animals at the outside of the food web, those that eat many other species but are fed upon by few. These are the top predators, and they receive the brunt of ingested toxins through bioaccumulation. They also show some of the most apparent signs of environmental poisoning: Published reports during the past half century have shown that at least 14 top predator species have had severe reproductive problems or population declines. This includes animals such as terns, gulls, eagles, otters, mink, and various predatory fishes.

One of the fish species, the lake trout, provides a classic example of the effects of pollutants on a predatory species. This fish was once an important commercial species. As far back as the 1870s, fishermen pulled more than a million pounds of lake trout yearly from Lake Ontario, the least productive of the lakes. Twenty years later, Lake Huron was yielding nearly 7.5 million pounds yearly and Lake Michigan about 3 million. But the lake trout's productivity, like that of other valuable commercial fish species, soon crashed. Overfishing accounted for some of the decline, as did destruction of habitat and the unintentional introduction of exotic species, such as the predatory lamprey, which attaches itself with its rasping mouth to the sides of the trout and feeds upon the living animal.

By the early 1960s, the lake trout had vanished from lakes Michigan and Ontario and was nearly gone from Huron. Its numbers had dropped to insignificant levels in the other lakes as well. The mid-1960s saw the initiation of plans to restore the Great Lakes fishery, including the lake trout. Since then, millions of dollars have been spent controlling lampreys and raising lake trout in hatcheries. Millions of lake trout have been released yearly in the Great Lakes, 2 million a year in Lake Michigan alone.

But to no avail. Only a few of the fish have spawned successfully, including some in Lake Superior, northern Lake Huron, and northeastern Lake Michigan. In Lake Ontario, researchers have found no evidence of naturally born yearling lake trout, though some eggs produced in the wild by released hatchery fish did produce fry. Southern Lake Michigan is also a failure as a spawning site. The lakes remain an unfriendly home for a healthy fishery. Almost all adult lake trout now inhabiting the lakes were captive-bred and released.

The cause of the reproductive failure apparently lies in the fish's own tissues. Lake trout feed on other fish and quickly accumulate large concentrations of toxins in their fats.

Mike Mac, a U.S. Fish and Wildlife Service biologist, is one of the modern-day Great Lakes explorers searching for the truth about what contaminants are doing to the natural world. "Way back in the middle sixties, when they started planting lake trout in Lake Michigan, they felt like this was eventually going to become a self-sustaining population," says Mac. "We started working on it in the middle seventies, wondering why 10 years or so after we started planting we were not seeing any young fish out in the lake. Because Lake Michigan notoriously had high levels of organic contaminants such as PCBs, we thought that it would make a possible reason why they are not reproducing."

Mac and the federal biologists with whom he works found the trout burdened with toxins such as PCBs and various pesticides. The toxins tend to concentrate in the fat and even turn up in the fats that females provide their eggs for food. The chemicals apparently do not do any noticeable damage to the adults, but they may affect the development of embryos and young fish. Some eggs simply do not hatch. In others, the newly liberated fry suffer developmental defects that cause them to swim erratically and eventually to die. In some, death does not come until the fry use up the yolk sacs that remain attached after hatching. This seems to occur at about the time of "swim up," when the young must swim to the surface to fill the swim bladder, a lunglike organ, with air to regulate buoyancy. Lake Michigan fry raised by Mac in the laboratory become sluggish at this point and never swim up. Instead, they slowly die off. This suggests that fats in the yolk sac flood the fry with contaminants that weaken and kill them.

Because lake trout are so heavily contaminated, it is impossible to determine which toxins are killing their young. However, the link between contaminants in general and the deaths of the fish seems solid. A Fish and Wildlife Service study of fish from three of the lakes showed that those from Lake Superior were least contaminated, those from Lake Michigan the worst, and those from Huron in the middle. When Fish and Wildlife Service biologists studied the survival rate of offspring from these fish, they reported that nearly all the eggs taken from Lake Superior hatched in captivity, while only 70 percent of fertilized eggs from Lake Michigan hatched. Fry survival rates were even more telling. About half the fry from Lake Superior and northwestern Lake Huron lived 139 days after hatching, while only 4 percent of the fry from Lake Michigan did so.

Because the evidence is indirect, many scientists are unwilling to point a finger and say that the failure of fry to survive should be blamed on toxins. But the correlation of the presence of toxins in adult tissues and in egg yolk with the inability of fry to live beyond a few weeks is certainly persuasive evidence of a link between contaminants and reproductive failure in the lake trout. A similar body of indirect evidence suggests that a variety of fish-eating animals are also failing to reproduce because of high levels of toxins in their tissues.

One such creature is the bald eagle, a species that barely survived the use of DDT and similar pesticides. After ingesting these toxins in the fish that are their main food, the eagles would lay eggs whose shells were depleted of calcium. This made the eggs weak-walled and easily broken, and eagle populations plummeted nationwide, as did populations of many other bird species, from ospreys and falcons to robins, in whose prey toxins occurred. These birds were saved primarily by the banning of many of the toxic pesticides early in the 1970s.

Since then, bald eagle populations have begun to spring back. But not along the shores of the Great Lakes. Hundreds that nest along the lake shores suffer reproductive failure even though the shores are prime nesting sites, offering a ready supply of the fish that constitute the bulk of the eagle's diet. According to Don Elsing, a biologist at Hiawatha National Forest in Wisconsin who is exploring the differences in reproductive rates between shore-nesting and inland-nesting eagles, those that nest within five miles of Lake Michigan breed normally for only two years and give up altogether after five years. In other areas, bald eagles breed for ten years or longer.

A team from Michigan State University, led by biologist Bill Bowerman, examines wild eagle hatchlings each year, reaching into nests to take blood samples. Tests show that lake-shore nestlings begin life with a major disadvantage—their blood is loaded with up to nine times more contaminants than is that of hatchlings from farther inland. Eggs show similarly high levels of contaminants. Abandoned eggs collected around Lake Michigan in 1986 showed 30.2 parts per million of DDE, a metabolite of DDT, and 55.3 parts per million of PCBs, while eggs taken the same year at Lake Huron showed 41.1 parts per million of DDE and 107.2 parts per million of PCBs. This contrasts sharply with an inland Minnesota egg taken in 1984 that showed only 1.2 parts per million of DDE and 1.9 parts per million of PCBs. We know that bald eagle eggshells lose 10 percent of their shell thickness at only 5 parts per million of DDE, and that a 100 percent loss of eggs results at 15 parts per million and higher. No one knows how high contaminant levels affect the young shoreline birds that do survive to maturity. Can they fend off disease, will they be strong enough to survive, can they reproduce?

Some clues to the answers to these questions are coming from laboratory research. Biologist Michael Fry, of the University of California at Davis, investigated the effect on California kestrels of the pesticide

Bill Bowerman and assistants band a juvenile bald eagle in Hiawatha National Forest in Michigan's Upper Peninsula. Studies show that those birds nesting within five miles of Lake Michigan are less successful at fledging young than are birds living farther from shore. Tests of unhatched eggs reveal high concentrations of toxics, passed on from the parents through their diet of lake fish.

dicofol, used to kill mites on fruit trees. He found that male birds became feminized, with testicles reduced in size or nearly lost. This was a physiological study that did not answer questions about how the changes affected kestrel behavior. But that aspect is being explored by Manon Bombardier, a young woman working on her master's degree in animal parasitology under David Bird, director of the McGill University Raptor Center outside Montreal. In her study, Bombardier dosed young male kestrels with one and two parts per million of the pesticide kelthane, which contains dicofol and is widely used on fruit trees in California, Texas, and New York. She found that young males treated with kelthane did not grow as large as untreated males. Because, among kestrels, larger birds dominate in social interactions, the treated birds were forced into submissive status. This has important implications for the reproductive success of the smaller birds. Male kestrels establish nesting territories for themselves and their mates. A submissive male will claim a smaller or substandard territory, making it a less effective breeder. Moreover, a smaller male that has never been dosed with pesticides will dominate a contaminated bird that might, had it never encountered pesticides, have been larger and more aggressive. Consequently, smaller, uncontaminated males that might under natural conditions have been left out of the breeding cycle can instead pass on their substandard genes to future generations, weakening the species.

Chilling indications of the hazards created by unleashed contaminants is surfacing in research conducted on Great Lakes gulls, terns, and cormorants as well. These birds and others, through bioaccumulation, can carry concentrations of toxins that are 25 million times greater than those that occur in Great Lakes water and 21 to 39 times greater than can be found in the fish the birds eat. When providing nourishment for the eggs they produce, female birds expel some of the pesticides and other contaminants they ingest. As in mammals and fish, the contaminants are in the birds' fat, and a portion of the female's fat is incorporated into egg yolk. Thus the young are the seg-

A graduate student holds one of the kestrels under study at McGill University's McDonald Raptor Center near Montreal, Canada. Researchers found that male birds dosed with pesticide were smaller and therefore less dominant than undosed males.

Dr. Jim Ludwig and his nephew, Matt, examine a deformed cormorant chick on Little Gull Island in North Green Bay, Wisconsin. The birds are veritable toxic sponges, accumulating concentrations of deadly chemicals nearly 40 times that found in the fish they eat.

ment of the population most affected by toxins, and the damage begins before they even leave their eggs.

Jim Ludwig is an independent biologist who has spent 30 years studying how contaminants affect the double-crested cormorant. He sees the cormorant as an *indicator species*, one sensitive enough to the harmful effects of toxins that its survival and condition can serve as a warning that contaminants are loose in an area. "They reflect the toxic problems embryonically, and they're quite sensitive at that state," says Ludwig. "Fifteen to 20 years ago we had such a DDT problem in the Great Lakes that a great many of these species, particularly sensitive ones like cormorants, simply couldn't reproduce at all. We had eggshell thinning problems that were severe. Cormorants laid eggs that were, on the average, about 25 percent lighter in eggshell thickness than they do today. As that problem cleared up, it allowed these more persistent substances to begin to express themselves."

Apparently, in the heyday of DDT, the inability of young in thin-shelled eggs to survive to hatching masked another problem: the appearance of toxin-induced birth defects that kill many hatchlings even if they live long enough to escape their eggs. One problem is edema in the tissues of the head, neck, and abdomen. "Edema," says Ludwig, "means that the animal has retained fluid, much like a pregnant woman does in her ankles, except that for a bird embryo that becomes very significant. What that does for a tiny bird like this in the confined space of an egg, it cannot pip its way out because it can't raise its head up to bring the egg tooth up against the wall of the egg. It swells up, it gets more and more confined, rotates the head down, and the animal dies."

Other birth defects also occur, including malformations in neck

bones that cause the neck to twist as much as 180 degrees and that ultimately lead to death—a condition common in terns but rare in cormorants. Cormorants are frequently born with defective beaks, with upper and lower bills twisted so badly that the birds die when they no longer have their parents to feed them. In some cases, the mandibles are so twisted that they grow back toward the skull, threatening to impale the bird should it grow older.

Cormorants around the Great Lakes are not being threatened with extinction by pollutants, at least not yet. They can be found in great numbers, having rebounded from a low of a few hundred birds in the 1960s, following their devastation by DDT, to the thousands of birds present today. The cormorant, says Ludwig, "is doing very well as a species and as a population. What we should be focusing on are the small species with a very high metabolic rate, like the Forster's tern, the common tern, that are much less able to compensate for the toxicity in their eggs because their reproductive rate is lower, their ability to replace eggs is lower. What I think is happening in the Great Lakes, perhaps elsewhere as well, is we're losing diversity. We're ending up with a whole lot of very resistant individuals, of things like ring-billed gulls, double-crested cormorants, and herring gulls, that resist the toxic contamination. The more elegant species, like the common tern, the Forster's tern, they're disappearing because they can't tolerate it. So you get a whole ecosystem shift."

Nowhere is that shift more dramatic than on Kidney Island, near the city of Green Bay. Built from mud dredged out of the Fox River, the island is mobbed by ring-billed gulls. But on one isolated corner is the last remaining Great Lakes colony of Forster's terns. They once nested in wetlands up and down the coast, but those wetlands are largely gone, and the tern is declining. University of Wisconsin biologist Tom Erdman has been studying the terns during the past decade, and his research reveals an insidious problem. "We have a population which appears to be stable if not slightly increasing," Erdman says. "But unfortunately, we're not seeing birds that we banded as youngsters. We're banding 300, 400, 500 young birds a year. We are not seeing banded adult birds."

The answer to just why the banded young are not reappearing in later years as adults may lie in their food habits. The thriving gulls scavenge for food all around the city of Green Bay, but the terns live only on fish caught in surrounding waters. Consequently, within weeks of hatching most of the birds have accumulated large doses of toxic chemicals. They begin wasting away.

The population on Kidney Island is growing, but not because the birds there are reproducing. It is expanding because of an influx from a cleaner colony about 40 miles inland. This bodes ill for the incoming birds. Says biologist Bud Harris of the University of Wisconsin, "What this really tells us is that Green Bay could be an ecological trap, because it's pulling in birds that are relatively uncontaminated. Once they get into this environment, they're going to begin to accumulate PCBs."

Bald eagles, cormorants, and Forster's terns are not the only birds affected. Other terns and some gull species are also touched by the

◀

The twisted and curved bill of a double-crested cormorant found in the Great Lakes is a sign of genetic monkey business. Many chicks are so deformed that they never make it out of the egg.

poisons. The effects on the young of these species include hazardous weight loss, immune suppression, behavioral changes, organ damage such as liver disease, and birth defects such as crossed bills, lack of jaws and/or skull bones, clubfeet, lack of feathers, twisted ankles, and dwarfed limbs.

These defects are a critical symptom of widespread systemic problems in the entire Great Lakes region. Mammals have suffered as well. Mink and river otters have almost vanished from some large sections of Great Lakes shoreline, especially within five miles of Lake Michigan shores. Two lines of evidence suggest that fish contaminants are the culprits here. The first is that captive mink fed a diet of 30 percent Lake Michigan salmon suffer reproductive failure and even death. The second is the fact that though muskrats live in the same habitat, they do well in areas where otters and mink cannot seem to survive. The muskrats pull through probably because they are primarily vegetarians and generally do not eat fish.

People are mammals, of course, and what is happening to the otters and mink is a warning that Great Lakes residents may want to heed. Says Jim Ludwig, "Forget about the birds. Ask yourself the question about human bioeffects. You know, I'm afraid to let my children eat fish from out of the lakes."

Gulls swarm around the grounds of the Niagara Power Project, Niagara Falls, New York. The subtle effects of toxic poisoning may leave us with huge populations of resistant species such as gulls while wiping out populations of sensitive species such as Forster's terns.

Water, Water, Everywhere, But Is It Safe to Drink: The Human Factor

Water can carry bacterial contaminants from human waste and various toxins recently dumped into it, but for the real dangers you are likely to encounter in a body of water you also need to look closely at the sediments and at water animals that you are likely to eat.

In the Great Lakes region, the greatest threat to human health is quite likely the eating of lake fish. It is becoming increasingly clear that although some fish species are making a comeback as lake waters are cleaned up, individual fish themselves are heavily contaminated. This alarming realization is dawning on us just as sport fishing in the Great Lakes region is recovering from its recent slump. Although fishing in the lakes has become a $4 billion-a-year industry, the fish being caught probably should never be eaten. A survey of Great Lakes fish by Wisconsin biologists revealed that one in four fish taken from the lakes is unsafe for consumption. The level of contamination varies from lake to lake. Lake trout from relatively pristine Lake Superior carry 4 parts per million of PCBs, while lake trout from Lake Michigan have 36 parts per million. For human consumption the maximum safe level of PCBs in fish is 2 parts per million.

At present, government figures suggest that because of contaminant levels, no one should eat more than 5 pounds of Great Lakes fish yearly. Unfortunately, sport anglers and their family members are, on

Another dead fish washes ashore at Roosevelt Beach on the New York segment of Lake Ontario. Bottom-dwelling fish and very large fish carry the heaviest loads of toxins. Anglers are warned not to eat more than 5 pounds of Great Lakes fish a year, but many people ignore the advice.

the average, each eating almost three times that amount. The important question, of course, is how exactly does this affect the human diners? We already know that in wildlife, the young are the most seriously threatened segment of the population. "Juvenile organisms are generally much more sensitive to the effects of contaminants than are adults," says John Giesy, a toxicologist with the University of Michigan. "Why is that so? Well, juvenile organisms are just literally developing from a single cell. That means, for instance, with a liver cell at that point, if there's an effect on the cell so it's defective at the time the primordial first liver cell forms, all subsequent cells have a very high probability of being affected, whereas in the adult if a few cells are damaged, they're simply replaced, and the effect on the organism is much less. So it's this critical window of time when the animal's at one cell, or two or four cells, that the probability of effect there is translated all the way through the entire organism's body and life."

A handful of studies suggests that what is true for wildlife species is true for humans, too. One study was done on laboratory rats. Since rat and human tissues are fundamentally the same, at least as far as biochemistry goes, the reaction of lab rats to contaminants gives us some clues about how contaminants may be affecting people. In an experiment at the State University of New York in Oswego, psychologist Helen Daly fed one group of rats a diet of 30 percent Lake Ontario salmon and compared their reactions to stress with those of a group fed uncontaminated Pacific Ocean salmon. Says Daly, "When life is pleasant it doesn't matter which diet they have, but as soon as we make life even mildly unpleasant, the rats on Lake Ontario salmon are hyperreactive to these negative events, or more reactive."

In one test, the two groups were taught to press a lever that released a food pellet into the rat cages. When the system was adjusted so that the rats had to press the bar 2, 4, 6, 8 times and so on to get a pellet, the rats fed uncontaminated Pacific salmon soon gave up. This is considered a normal response. The rats fed Lake Ontario fish took a different approach, one that might be called obsessive, or at best excessive. They overreacted and pumped the bar thousands of times for up to an hour before stopping.

Daly subsequently tested the offspring of rats fed Lake Ontario salmon. Even though the young themselves had had no contact with the contaminated food, they still overreacted in stressful situations. "And when we retested these babies in adulthood," says Daly, "they showed the same kinds of results."

This study suggests that food contaminants could have harmful effects on humans as well. And that is precisely what was shown by a series of tests on the offspring of nearly 250 Michigan women who regularly ate fish from Lake Michigan—something roughly equal to two or three lake trout, coho salmon, or chinook salmon monthly for at least six years before pregnancy. Compared to children from women who did not eat fish, the fish-eating women's offspring were slightly smaller and less responsive. Joseph Jacobson, one of the psychologists at Wayne State University who conducted the study, says, "One thing that surprised us a little was that it was the level of PCBs in the umbilical-cord blood that predicted poor performance on this test. It didn't seem to matter how much PCBs the infant had gotten from nursing. And this was contrary to what had been expected. There's so much more PCBs in the breast milk and infants are exposed at so much higher levels that, until that time, people had pretty much assumed that it was the nursing that was the problem. These data show that that doesn't seem to be the case, that it's the prenatal exposure, how much crosses the placenta when the fetus is still vulnerable.''

The harmful effects apparently are long lasting. Jacobson and his wife, Sandra, tested the infants again at age four to see if the youngsters whose mothers ate contaminated fish had caught up with the control group. They had not. Joseph Jacobson explains: ''We gave the children an extensive battery of psychological tests, and what we saw was that there was a particular domain where these children seemed to be having problems, and it had to do with short-term memory, or attention. We have a computer game which showed pictures, and we tested the child's ability to remember the pictures. And on those tests, as we had seen with the infants, short-term memory seemed to be affected.''

''It's important to understand that the children who are more exposed to PCBs are not retarded,'' says Sandra Jacobson. ''They're really performing in the normal range. But what we're talking about is diminished potential.''

A gondola offers Toronto tourists a little piece of paradise on Lake Ontario. Below the sweet-water surface is a black history of abuse. Once, no one believed that the vast Great Lakes could be killed. Now the question is whether they can be resurrected.

The women who took part in the Jacobson's study came from towns around Lake Michigan. They represent as many as 100,000 other women along the shoreline. The subtle but significant effects uncovered in the study could be widespread among Michigan's next generation. Says biologist Jim Ludwig, ''What has occurred basically is that some users of these [contaminating] materials transferred the cost of disposal to the wildlife and to unborn generations of human beings in these systems by disposing of them in an unsound, unsafe

manner. They dumped them into the water, and they got them into the sediments. The question is, do we want to allow that to continue? It is an ethical question, and it will continue even though there isn't material now coming out of the pipes, because it's in the sediment, and the sediment can give the material back to the water."

All of which leads to the critical question: With the Great Lakes so thoroughly despoiled, can they be cleaned?

Cleaning Up Our Act

Sometimes you have to wonder if anything as vast as the Great Lakes, once they become polluted, can ever be restored. This seems especially doubtful when you consider not only the economics of the situation—all those factories pumping toxins into air and water, and at the same time providing jobs and adding to the gross national product—but also the magnitude of the sources of pollution. All those cars—more than 137 million—billowing toxins into the air. Will enough of us be willing to give them up? Will the automotive industry do what is necessary to reduce automobile pollution? And will any politician take on the challenge of cutting into business for the sake of a better quality of life for our own and future generations? Take one look at environmental history, and you can't help but wonder if anything could seem more unlikely.

Nevertheless, some people believe we can do the job. One of the leaders in this field is the International Joint Commission. In its Fifth Biennial Report on Great Lakes Water Quality, the commission made several recommendations for cleaning up the lakes. It urged the U.S. and Canadian governments to stop the inflow of persistent toxins into the Great Lakes, to give priority to actions that protect and restore the Great Lakes ecosystem, to prepare comprehensive public information and education programs on the dangers of lake pollution, to develop and implement plans for restoration of highly polluted areas, and to develop means for the prevention and containment of toxic spills from vessels and other sources.

Meanwhile, as government officials at various levels try to decide how to tackle these recommendations, concerned citizens are initiating cleanup campaigns on a local level. A citizens group is trying to remove garbage from the Buffalo River, which lies between lakes Ontario and Erie just above Niagara Falls and until recently was billed as one of the most polluted rivers in the nation. A citizens task force is also at work trying to clean up the Grand Calumet River, which the Environmental Protection Agency has targeted as a national-priority cleanup site. Groups such as the Friends of the Buffalo River are also helping to develop the comprehensive restoration plans called for by the Joint Commission.

But too often the problem is too great for citizens' groups and perhaps even for national governments. What shall we do about the toxic sediments that underlie rivers even after the surface garbage—the tires and oil drums and other debris—have been removed? What will we do about toxins that rush in on the wind from hundreds of miles away? What will we do about fish whose meat has become deadly? How can we clean up all the sediments at hundreds of sites around the lakes? Can we stop the toxins that leach from dump sites near Buffalo, New York, poisoning Lake Ontario? Is our belief that we can right these wrongs just another symptom of the same foolish arrogance that led us to pollute the lakes and the winds in the first place?

We call the birds and other animals we have poisoned *indicator species* because, in their sensitivity to environmental degradation, they indicate to us that something is amiss in the homeland, in the habitat that we share with them. We say they are an early warning to us, a sign that we need to take action before it is too late. But are the deformed birds and vanished mink and otters really indications that things have already gone too far? By the time we notice their warnings, is it already too late to find a sure cure? Hasn't the integrity of the entire ecosystem already been ruined by the time bald eagles and Forster's terns and ring-billed gulls and double-crested cormorants and mink and common terns and snapping turtles and lake trout and beluga whales start showing signs that something is wrong?

Perhaps it is not too late to correct the situation, but already the damage stretches far into the future. The International Joint Commission may have summarized the situation best, or at least the most succinctly: "What our generation has failed to realize is that, what we are doing to the Great Lakes, we are doing to ourselves and to our children." Indeed, the cleaning of the Great Lakes will take so many years, so many decades, that should it ever come about, it is not likely that any of us—or even, perhaps, our children—will be there to know whether it was possible. Perhaps the legacy to be gained from this, should we be willing to observe it, is that if we are capable of destroying so vast an ecosystem as the Great Lakes then we are capable of bringing ruin to anything on the globe. We are a consummately dangerous species and should, for the better protection of the globe, treat ourselves like one. We should learn to be wary of ourselves, of our goals, our inclinations, and our abilities. We should learn to restrain our seemingly infinite ability to create technological miracles, because technology has itself so often turned into a problem. We should act with the caution born of a reasonable distrust. Until such a time as we learn these things, we will continue the course of negligence that turned the Great Lakes into a human health threat, and we will ultimately jeopardize the environmental integrity of the globe.

HATARI, WATEMBO: AFRICA'S ELEPHANTS AT RISK

S ome 2,500 years ago a fleet of ships with curved prows and single sails moved through the Pillars of Hercules to brave the harsh waters of the Atlantic. The men on board knew well that they were leaving behind the gentle seas of the Mediterranean for the devil waters of the ocean. It was an era in which ships were poorly equipped for human comfort. Because rowers had to sleep at night on the benches where by day they heaved at oars, they preferred to come ashore each night to camp. But to cling to Atlantic shores was to risk being shattered upon jagged rocks, while the alternative—moving out to sea—was just as bad. The prevailing winds could push a fleet far out from land, hopeless of return.

Nevertheless, this fleet of Carthaginian vessels braved the violent sea. They came from a city of North Africa that rivaled Rome for dominance of the Mediterranean, and they were in search of places where they could establish settlements and extend their influence.

They sailed along Africa's western coast, moving south to the Tensift River, near the Atlas Mountains. Presumably they furled their sail there and rowed up the river, looking for a promised land. What they found was something that we will never see again in that region, for the ensuing years robbed the world of the sight. It was a vast number of elephants feeding in a marsh—as the Carthaginian Hanno recorded the sight, a marsh "haunted by elephants and multitudes of other grazing beasts."

Hanno's account, written in 480 B.C. and preserved only in Greek translation, marks the first known reference to the African elephant. The elephants of that region are long gone, wiped out by ivory hunters in behalf of the Roman Empire. Their demise was, in a sense, the beginning of a long, slow death for the African elephant, the throes of which have continued into our time. Since 1980, poaching for ivory has cut the number of the world's remaining African elephants by half.

Not long after the Carthaginian voyage, elephants became much better known to the ancient world, and the exploitation of the species began. In the fourth century B.C., Alexander the Great took his troops east, plundering and pillaging from his native Greece all the way to India. Along the way he encountered a weapon he had never seen before. It was a vast gray creature with a long nose that dragged on the ground, ivory tusks, and a piercing, trumpeting voice: the elephant.

The elephants that Alexander met were trained for battle and decked out in armor. Against men and horses who had never seen them before, they were nearly invincible. Their trumpeting terrified warrior and horse alike. The mere sight and smell of them was enough to rout an army. Naturally, Alexander could not fail to requisition a few elephants for himself. He used them for both combat and cargo,

◀ A big bull elephant feeds after a rain shower in southeast Kenya. His rust coloring comes from bathing in the region's red-orange soil.

presumably loading them with the booty of the nations from which the elephants came.

Not long after Alexander's day, the Egyptians started importing elephants from Asia. Both the Egyptians and the Carthaginians turned to capturing wild African elephants as well. Eventually, Carthage would win enduring fame for its use of elephants during the war against Rome, when Hannibal landed on the Iberian Peninsula in the third century B.C. with an array of war elephants. He used rafts to float the elephants across rivers and led them on a long march through the Alps before engaging Roman troops in a war he would eventually lose.

At first the elephants were terrifying in battle. Towers were erected on their backs from which archers could shoot and throw spears. The animals were armored, with spears sometimes tied to their tusks. But they worked best against men and horses who had never encountered them before. After a little experience, warriors soon learned that they could outmaneuver the lumbering beasts. Lightly armed soldiers would fend off the behemoths with javelins, then spear the fleeing animals in the soft areas under their tails. At this point the elephants would begin to panic, spreading discord and panic among anyone who got in their way, which, if the elephants were retreating, usually meant their own men. For this reason, the Greeks called them "common enemies." Pliny, in *The War with Hannibal*, gave an illustrative account of a battle in which a Roman officer led his men

> into the very thick of the turmoil caused by the solid mass of elephants and ordered them to let fly with their javelins. Every weapon found a mark—and indeed beasts of such size and packed so closely together presented no difficult target; but not all were hit, and those which had spears sticking in their backs turned and ran, like the untrustworthy beasts they are, and in their efforts to escape carried with them the others who were still untouched. . . . The poor brutes charged their own masters and caused even greater carnage amongst them than they had caused amongst the enemy—inevitably, because a frightened beast is driven more fiercely by terror than when he is under the control of his rider.

To prevent such loss of control, riders were given special equipment for establishing order in a disorderly elephant—a mallet and a carpenter's chisel. If an elephant ran amok, the rider would center the chisel between the elephant's ears where head and neck met and hammer it with a heavy blow. It is not clear how the rider survived the sudden collapse of the dying multi-ton animal upon which he rode.

Elephants were also used in the ancient world for entertainment at various circuses and games. Sometimes trained elephants performed tricks. In one case, elephants dressed in flower garlands danced and then sat down to eat at a banquet table. To the delight of the audience, the elephants' decorum soon broke down as the animals sprayed each other and their trainer with water.

Less enchanting were the staged fights that pitted elephants against gladiators or other animals. Julius Caesar's rival, General Pompey, once sponsored a circus in which 18 elephants fought to the death against men armed with spears. One elephant, stabbed in the feet, dragged itself along on its knees, throwing gladiators into the air like dolls. The men spun through the air, whirling around and giving the audience quite a laugh. But if the Roman spectators of the time could look with gaiety upon the sight of a man pinwheeling aloft, they took a dimmer view of the elephants' anguish. When the animals panicked, running around the arena and trumpeting in what seemed desperation, the crowd pitied them and cursed Pompey for his tastelessness in creating the spectacle.

A worse fate than the gladiatorial ring awaited most of the elephants of northern Africa. Rome, Greece, Egypt, and Middle Eastern nations all plundered wild herds for ivory. Ivory was used in statues and furniture and for luxury items such as combs, boxes, seals, bird cages, thrones, scabbards, brooches, book covers, scepters, musical instruments, chariots, and even such household items as doors. By the first century A.D., Roman demand created a shortage in African ivory supplies. By the seventh century the elephant was apparently extinct in North Africa, and Rome was forced to turn to Ethiopia and Somalia for ivory.

With the collapse of Rome in the fifth century, European trade in ivory dropped off. Nevertheless, demand was still in the air. Arabia was a big ivory market, drawing from Asia. Much of the demand was met by fossil ivory taken from the exhumed remains of mammoths and mastodons, distant prehistoric relatives of today's elephants. For a long time the elephant herds of Africa south of the Sahara were relatively untouched by the markets of the Northern Hemisphere, save perhaps along Arabian trade routes that cut through East Africa.

By the seventeenth century, the Arabs dominated the ivory business, and Europeans became ambitious to have a share of the enterprise. Consequently, the yen for ivory was one impetus for a Portuguese expedition along the African coast at the close of the fifteenth century. Led by Vasco da Gama, the sailing fleet rounded the Cape of Good Hope, the first European vessels to do so. In the next cen-

THE AFRICAN ELEPHANT

THE ISSUE African elephant populations have declined drastically—by as much as 95 percent in some areas—during the past decade.

THE CAUSE The killing of elephants by poachers to supply the ivory markets primarily of Japan, the United States, and Europe. Elephants are also jeopardized by loss of habitat as agriculture reaches deeper into the African wilderness.

EFFECTS ON WILDLIFE The elephants themselves have been reduced about 50 percent continentwide. In some local areas and some individual nations they have been practically wiped out. The elephant's complex social structure has been so badly damaged in some areas that it is not clear whether surviving elephants will be able to sustain their populations through reproduction. Other species may be affected because the elephant is a keystone species, one that shapes its environment. The changes that the elephant brings to its habitat are beneficial to a variety of other savanna species.

EFFECTS ON HUMANS Loss of the elephant could have a drastic impact on the tourist businesses of several nations, including Kenya and Tanzania. The danger posed by poachers armed with modern firearms has at times jeopardized the tourist trade. Conflicts between the interests of local residents and those of elephants may affect how land is used and who may use it.

EFFECTS ON HABITAT Unclear. Loss of the elephant may affect the rapidity with which trees populate savannas emptied of elephants, potentially changing the types of species that can use the land.

tury—a time of worldwide exploration that would usher in an era of settlement such as history may never witness again—Portugal dominated world trade and established colonies along the African coast.

The Portuguese were forced to restrict their settlement of Africa to the coast because the rush for colonies and trade had left the government overextended. It had neither the funds nor the troops to set up and defend colonies in the African interior. But the Dutch, who started making inroads on the southern Africa coast around 1650, soon did move inland. This, too, was against their government's will, so the Dutch who settled the interior were not part of an official colonial policy and were therefore on their own. But many settlers accepted those terms, wresting cropland from native peoples without the support of troops. During the first 70 years or so of Dutch colonization, they wiped out a wide range of species in large parts of southern Africa, including lions, hippos, and elephants. Nevertheless, by the beginning of the nineteenth century, most of the interior was still unsettled.

By the early 1800s, however, the English took over South Africa from the Dutch as a by-product of the Napoleonic wars and began moving inexorably northward into the interior. By the middle of the century, the British, French, and Germans were also cutting into Africa from the east, where the Arabs had long dominated trade.

At this juncture, several things happened that brought the elephant back into European consciousness and gunsights. Industrialization had created a new financial order in Europe and North America. A rising middle class craved the accoutrements of wealth and culture. Among the most prized of these accoutrements—major fads of the time—were billiard tables and pianos. One demanded ivory for balls and the other for keys.

This demand alone might not have been enough, however, to inspire widespread hunting of African elephants. For a long time Africa was simply out of reach, protected by distance, the threat of disease, and the hostility of native peoples. But in the nineteenth century these obstacles were overcome. The invention of the steamship and the opening of the Suez Canal took care of the distance problem, quinine took care of malaria—the primary disease barrier—and the invention of the machine gun took care of recalcitrant native residents. Demand for ivory and access to it thus coincided as they never had before.

By the end of the nineteenth century, ivory hunting was already a threat to elephant survival. Though elephants still roamed large parts of Africa, in some areas ivory hunting had wiped them out. It became clear that as hunting reached deeper and deeper into the continent, elephants would vanish everywhere. In 1894, the government of South Africa's Transvaal Republic, at the persistent urging of Paul Kruger, its president, created Africa's first wildlife sanctuary, Pongola Reserve. In 1895, the Sabi and Shingwedzie reserves were created, also in South Africa, protecting several thousand square miles of country that had been hard hit by hunters. In the whole area, only 10 elephants remained. In 1926 the two reserves were combined into 7,000-square-mile Kruger National Park, and today, thanks to strict protection, the park apparently has all the elephants it can hold.

Although men such as Kruger a century ago had enough foresight to plan protection for beleaguered elephant herds, little was known about the behavior and needs of elephants. The idea that the survival of elephants might be closely linked to the survival of the many species with which they shared their range was alien to the naturalists of the time. But in recent years—since the 1960s—biologists have discovered an important link between elephants and other wildlife. The elephant is a *keystone species,* one that helps shape and maintain the character of its habitat, one crucial to the survival of many other species that share the elephant's habitat. How the demise of the elephant would affect its habitat is as hard to pin down, but it is likely that loss of the elephant would be devastating to many other species.

Ironically, just as we were learning this and other important information about elephant behavior and ecology we were also witnessing the culmination of an ages-long scenario—what looked to be the final destruction of the elephant for ivory. Since 1980, the ivory trade has destroyed half of Africa's remaining 1.5 million elephants, which were themselves a tiny remnant of the 10 million estimated to have roamed the continent in the 1930s. Had ivoryhunting gone unabated, the elephant would have been lost by the end of the twentieth century. Fortunately, illegal ivory hunting in most areas has been very nearly brought to an end. Nonetheless, because of the extent to which elephant populations were reduced, and the intricate social behavior that governs the elephant's life, the poaching of the 1980s is still likely to have dreadful effects on surviving elephants.

A forest elephant takes a drink at a clearing in the Central African Republic. The elephants visit open salt flats daily to dabble in the mud, quench their thirst, and eat mineral-rich soils.

Part of a herd of some 150 elephants gathers at Kenya's Amboseli National Park. Such large groups present a tempting target to poachers, who can earn a year's income from just one of the big tuskers.

◀

Elephant researcher Iain Douglas-Hamilton, at right, explains the intricacies of field study to a team of British journalists. His pioneering work led to methods of individual elephant identification that have become standards in the field.

Elephant Behavior and Habitat

Two types of elephants roam Africa. The best known is the bush or savanna elephant. The other is the smaller elephant of Africa's rainforests. The forest elephant grows to a maximum height at the shoulder of about 8 feet, while the largest bush elephant on record exceeded 13 feet by a couple inches. The forest elephant also looks different from the bush subspecies. Its ears are smaller, its back is rounded rather than swayed as in the bush elephant, and its tusks are long, thin, straight, and pinkish. The tusks are also harder than the bush elephant's.

The forest elephant is hard to observe. The forest in which it lives is so dense that individual elephants can be impossible to see when only 10 yards away. Even trying to determine how many forest elephants live in a given area requires use of some indirect methods. Biologists count the number of elephant droppings and apply a special formula to come up with a population estimate.

Nevertheless, information on the forest elephant will soon be available. Biologists working for the New York Zoological Society's research department, Wildlife Conservation International, are studying forest elephants now in the Congo, Zaire, and the Central African Republic and will soon provide information on the subspecies. Meanwhile, we know next to nothing about the forest elephant except that its numbers seem to be declining as the ivory trade comes to Africa's rainforests.

Most of what is known about African elephants concerns the bush subspecies, and what has been learned about them is fascinating. The first study of living wild elephants was begun in 1965 at Tanzania's Manyara National Park by British biologist Iain Douglas-Hamilton. Because he needed to identify individual elephants in order to make sense of the animals' social behavior, Douglas-Hamilton soon developed a research technique that has proved indispensable to elephant biologists ever since. First, he learned that you can identify individual elephants by their ears. Ears are distinct because of differences in size, shape, scars, holes, and tears. Even more important, the veins of the ears of each elephant form patterns that are as distinctive and unalterable as fingerprints are in humans. Once Douglas-Hamilton learned that he could recognize individual elephants by their ears, he developed a photo-file system that eventually allowed him to catalog nearly 500 Manyara elephants. He did this by photographing the ears of each elephant in each group that he studied. This was not easy. He wanted ears from right, left, and full frontal. To get full frontal photos of the ears required that the elephant look straight into the camera with ears extended. Elephants usually strike this pose only when threatening whatever is in front of them with a precharge display, a situation that often compelled Douglas-Hamilton to engage in some swift tree climbing to escape an irate elephant.

Douglas-Hamilton's file enabled him to follow individual elephants and observe how they relate to one another and use their habitat. His technique has been adopted by other researchers, notably Cynthia Moss, who started working with Douglas-Hamilton in 1968 and then, four years later, launched her own elephant research in Kenya's Amboseli National Park.

Douglas-Hamilton, Moss, and others who joined in their studies are still in the field, gathering data on the African elephant. It is no exaggeration to say that all of what follows flows directly from their work.

Elephants are highly social animals. As a rule, they live in groups of about 10 animals, though group size varies depending on the amount of food the habitat can provide. The groups are led by matriarchs in their forties or older. Years of experience have taught them where to find water and food during each part of the year. Matriarchs set the time and course of travel and are the rallying point for the group whenever danger threatens. You may have seen a film of a matriarch defending her family group, her eyes fast on the camera, trumpeting with head raised and ears extended, shuffling about in a cloud of dust, swaying with pent-up energy even when she was standing still. In your imagination you may have thought this was a herd bull, but in fact it is much more likely to have been a herd mother, a matriarch.

The group is a family in the full sense of the word. All the individuals in the group in one way or another grew up together. Chances are that not a single one of them has ever been alone. They are one another's constant companions. In addition to the matriarch they include the matriarch's sons, daughters, grandsons, and granddaughters and probably some sisters, nieces, and nephews. They are intimately attuned to one another. They feed together, wash, drink, and rest together. Older females help rear younger calves, regardless of who is the mother. Mothers and offspring, however, are tightly bonded. During the first months of life, mothers and offspring are rarely more than a few feet apart, and they frequently touch.

If group members do become visually separated, they remain in vocal contact. This is relatively easy for them to do because elephants produce low-frequency vocalizations called infrasounds. Many of these calls are at frequencies so low that the human ear cannot detect them. However, the human body sometimes can. If you stand beside an elephant that is issuing one of these low calls, you might feel the vibration of the noise against you, much as you might feel the impact of a loud blast of thunder. Because these signals are made at such low frequencies, they can carry for several miles. This may explain how elephant groups separated by five or six miles can still coordinate their movements. It also means that an individual standing by itself, a few miles from other elephants, may nevertheless still be a functioning part of a group. "Alone" may have an entirely different meaning to an elephant than it does to a human.

The oldest elephants in any given family group, under natural conditions, are in their sixties. Much of their lives had already been lived before the first elephant study was initiated about 25 years ago. Consequently, biologists have not yet determined how the oldest elephants in different family groups might be interrelated. It is known that individuals from different matriarchical groups often greet one another and that matriarchical groups sometimes combine to feed and travel together, with the youngsters in each group playing together. It is not unlikely that the matriarchs of such groups are related—mother and daughter, perhaps, or sisters. In any event, bonds clearly exist between various family groups, so the large, combined units are called *bond groups*.

The life story of the average female, or cow, elephant goes something like this. She weighs about 250 pounds at birth and stands about three feet tall. Although lacking the tusks of the adult, she is a nearly perfect miniature of a mature cow. Her brain is about a third of its adult weight of 10 pounds, so she has a lot of developing, growing, and learning to do. She cannot survive without her family. For the first 10 years of life she is a mere child. She will not mature until she is nearly 20 years old. She is closely watched by adults—during the first months of her life she will spend 99 percent of her time with her mother, but even in later months she will be guarded by mother, older sisters, and aunts. A simple alarm cry will bring the whole herd to her defense.

She begins to use her trunk when about a week old, touching everything around her, exploring ceaselessly. Clumsy at first, by about three months she begins spending more time handling—or perhaps we should say "trunkling"—vegetation and begins to eat grass rather than dieting exclusively on milk. She may kneel to eat grass directly, just as she may kneel to drink water. At four months, though, she begins to experiment with sucking water into her trunk and spraying it into her mouth. By nine months she is rather adept at picking up grass and spends about half her time foraging. She still nurses at the two breasts between her mother's forelegs, and she sometimes takes food from the mouths of other elephants or eats food they drop. In this way she learns what she will eat when she is weaned.

She plays with other calves, too. This is important, because during play she and other calves establish lifelong bonds and learn skills they will use later in life. Bulls, for example, spar with one another, unknowingly preparing themselves for the battles they may fight later when they are in breeding condition. They will also establish a sort of hierarchy that determines who is dominant and who is not, a social structure that can help them to avoid fights when, in later years, they are bigger and more dangerous to one another.

The calf's permanent tusks begin to grow when she is about two years old. The first set were milk teeth that fell out when she was a yearling. The permanent tusks are modified incisors made of dentine tipped with enamel. Once they start growing they never stop. They could grow as long as 16 feet in a female, 20 in a male, but they rarely reach such lengths because daily use wears them down. They are used to knock over trees and chip off bark, to fight, and to dig. Each elephant is right- or left-tusked, so one tusk is usually more worn than the other.

The calf reaches sexual maturity at 11 or 12 years old, though she is still only about two-thirds of her adult height of 8 feet and weighs considerably less than the 6,000 pounds of a mature cow. If a breeding bull came into her group, she would probably flee him, but if he succeeded in mating with her she would have her first calf two years later. Two years after that she would be ready to breed again and would, potentially at least, give birth every four years until she was about 50. By then her birth rate would slow and any calves she had might fail to survive. At the end of that decade she would enter old age. She might even be a matriarch herself, drawing upon her years of experience to lead the herd to food, water, and safety. But her days would be numbered. In her late sixties her teeth would finally wear out, she would be unable to eat, and her life would ebb quietly away.

The cow's entire life would have been wrapped up in her family group. This is in marked contrast to the life of a male calf. He would live much as the female did until he, too, reached sexual maturity at 11 or 12. By his twentieth year—but usually within a couple of years after sexual maturity—he leaves the family group. He might stay close by and even follow the movements of his maternal group, but he would do so in the company of other young bulls. In time, though, he would join an all-male herd that probably lives in an area removed from habitat used by maternal groups. In the years ahead he would learn the dominance hierarchy of the bulls in his region and from time to time attempt to breed with available females.

In breeding he is not likely to be much of a success. Dominance in bulls is based on size, and since elephants grow throughout their lives, the oldest bulls are the biggest. The big males generally thwart the breeding attempts of the young, and besides, the females don't like younger bulls. They generally refuse the young bulls' entreaties. Females prefer mates over 40 years old and, biologically, this makes good sense. A bull that old is a proven survivor and is likely to provide the best genetic inheritance to the offspring.

At around 30 years old, bulls enter a physiological state called *musth*. In old bulls, the occurrence of musth is fairly predictable, arriving at a set time of year and lasting for two to five months. In bulls 25 to 30, for whom musth is a new development, the condition is sporadic, lasting only a few days to a few weeks.

The word *musth* is Indian in origin and means "aggressive." And aggressive the musth bulls definitely are. During musth, a bull's blood testosterone levels are high, heightening his sex drive and his hostility toward other males. A musth bull is eager to fight if fighting means he might win a female, so he is particularly dangerous to other bulls. Nonmusth bulls, regardless of size and rank, usually defer to musth bulls. Pulling rank, apparently, is simply not worth the hours-long fight that might ensue if a big nonmusth male decided to knock the chip off the shoulder of the musth bull. Besides, in confrontations with nonmusth bulls, bulls in musth generally win, regardless of size.

Musth bulls, fortunately for elephants that have to put up with them, show several telltale signs of their condition. The musth bull's temporal glands, on the sides of the head, swell and send a fluid secretion streaming down the face. He dribbles urine constantly as he travels about, advertising his condition and location. The odors of an old musth bull's urine and temporal gland secretions are enough to cause some young bulls to go out of musth. The odors also excite females, which trumpet and call shrilly when they encounter a musth bull's trail.

The musth bull also uses visual cues. He carries his head tucked in, chin to chest, and flaps his ears in a characteristic way. His stride is longer, his walk seems more purposeful, and he issues a distinct, low vocalization that is a love song to lustful cows and a warning to competitive bulls.

Young musth bulls do not lose all sense of discretion. They still tend to defer to older, bigger musth bulls. And the old bulls of a given area generally avoid fighting one another by entering musth at different times or by frequenting different areas.

When a musth bull finds a maternal herd, he searches for cows in breeding condition and mates with them, staying with each cow for two or three days. During that time he will keep other bulls from harassing the cow, giving her a rest from the ministrations of numerous eager young bulls.

As a species, the African elephant plays a big, and as yet poorly understood, role in the ecology of its habitat. It functions as a keystone species, essentially creating its own habitat. Elephants shape their environment as a by-product of eating. They require vast quantities of food, putting away up to 770 pounds of vegetable food and 40 gallons of water daily. Their food is a mix of grasses and browse from bushes and trees. The proportion of grass to browse varies with the seasons. They graze for grass primarily during wet seasons, turning to browse during dry seasons and droughts, when grasses decline.

The sheer quantity of food and water they take from their habitat has far-reaching effects on the habitat itself and hence on other spe-

cies. During droughts, a single elephant may knock down 1,000 trees in a single year. This can clear an area of trees and keep the land open for hoofed grazing species, such as antelope and zebra, which in turn feeds various predators. In some areas, elephants also help ensure that new trees replace those lost, so that a cycle involving the destruction of mature trees and the rebirth of new trees goes on endlessly. In Ghana, biologists found that the seeds of many fruits eaten by elephants pass unharmed through the elephants' digestive tracts. By eating fruits full of seeds elephants helped disperse at least 11 different tree species throughout one park in Ghana. Seeds from three of the species would not even germinate without elephant dung, and in other species the dung helped the seeds grow faster.

Another study showed that about 75 percent of acacia seed pods that pass through an elephant sprout successfully, compared to only 12 percent for seeds untouched by elephants. It seems that the digestive process softens the seeds' hard outer coverings and the dung acts as fertilizer.

Elephants are also helpful to other species during droughts because they will dig into river beds, creating ditches that fill with underground water. During severe droughts these ditches may be the only sources of standing water for many miles and may be instrumental in the survival of several hoofed species as well as some monkeys and predators.

Oddly enough, considering that it is an apparently immutable biological phenomenon, the elephant's role as a keystone species is not without controversy. Some wildlife managers see the elephant's leveling of woodlands as harmful. They believe that elephants are destroying natural woodlands, not merely suppressing tree growth as part of a process that creates a habitat hospitable to many species. They believe the knocking down of hundreds of trees is not part of a natural cycle, but was imposed on the elephant and Africa's woodlands by confining elephants to reserved lands. Under natural conditions, these managers say, the elephants would migrate to other feeding areas as grasses became scarce or at the very least would not need to concentrate their tree-crushing activities to limited areas. But as lands outside the reserves have been given over to farming, the elephants have been locked into relatively small areas. As a consequence of the limited amount of living space, the elephant's voracious feeding habits seem to threaten woodlands with ruin. Some studies suggest that the ebbing and spreading of woodlands is a natural cycle that involves the elephant, but in nations where wildlife management is dominated by administrators opposed to this viewpoint a special technique has been developed for protecting trees.

This technique is called *culling*. Culling involves the purposeful killing off of a segment of a wildlife population to reduce its numbers. This principle, in fact, underlies many of the fundamental tenets of wildlife management. Deer seasons in the United States are a form of culling, though done by amateur hunters. In Africa, culling often is performed by professional hunters. Culling elephants is a relatively easy task if modern weapons are used. Professionals with automatic weapons creep close to a maternal group and make a few small noises

◀
A large bull, temporal glands streaming, tests an acacia tree's stability. During a drought year, an elephant may down a thousand trees for food. Hooved grazers, like zebras and wildebeests, may benefit from elephants' tree-terrorizing ways, since fewer trees means more open grassland.

to alarm the animals, which then congregate near the matriarch for protection while trying to determine the source of the noise. Once the animals are grouped, an experienced hunter can down an entire family of 10 or 12 animals in roughly 30 seconds. The ivory and hides can be sold, and meat can be either sold or distributed to local people. Even the elephants' feet may net a profit if fashioned into wastebaskets or stools.

Culling divides even elephant biologists into pro- and anti-culling camps. Many biologists are disturbed by the idea that the species that brought us global warming and ozone depletion is now going to take upon itself responsibility for balancing the African ecosystem. They believe that we know too little about elephant ecology and family relationships to be certain that culls are not harmful in the long run not only to elephants but also to their habitat and to other species.

Those who favor culls are convinced that woodland destruction by elephants will have harmful effects not only on the habitat, but also on all the wildlife species dependent upon woodlands. But cull opponents respond that the declines in woodlands brought about by elephants are part of a natural cycle of ebb and flow, with elephants increasing as woodlands decline and declining as woodlands regrow. As Brian Walker, of the Center for Resource Ecology in Johannesburg, South Africa, wrote in *Problems in Management of Locally Abundant Wild Mammals,*

> Very often, many of our management actions which are aimed at conserving ecosystems and species are the opposite of what we should do. We cull animals when they increase, and we save them or introduce more when they decrease, to keep things constant. We spread water supplies evenly all over the reserves to remove spatial variation, and we use patch burning for the same reason. Most of our actions are in fact aimed at preventing the variability in time and space which in all probability is essential for the maintenance of ecosystem resilience.

In addition, research showing how elephants help the dispersal of some tree species suggests that the relationship between elephants and woodlands is not as simple as cull proponents believe.

Culling operations themselves do little to inspire confidence in skeptics. In the mid-1960s, for example, the wildebeest, a hoofed animal, was increasing during a dry period in South Africa's Kruger National Park. Since the habitat seemed to be declining, a wildebeest cull was started in 1965. In 1971 a wet period began, and two years later culling was stopped because wildebeest numbers seemed to be dropping too fast. Even after the culling ended, the population continued to decline. Finally, in 1979, the park started culling lions to protect the failing wildebeest. Wildlife managers finally concluded that wildebeest tend to do well during dry seasons, when grass is short and of high quality. Their numbers then increase. When wet seasons arrive, however, they tend to decline, in part because lions increase fourfold. The explosion in wildebeest populations during a dry season apparently offsets the decline that occurs during the debilitating wet seasons.

In elephants, the wiping out of entire maternal herds may hold special dangers for the populations of a given area, since each herd may

▶

A ranger patrols Tsavo East National Park in Kenya with an automatic rifle. Tsavo is considered the front line in the war against illegal hunting. The Kenya Wildlife Department has increased funding and enforcement of its antipoaching laws.

represent a distinct genetic line. Destroying a whole genetic line may have harmful effects that have not been recognized simply because no one has studied this possibility. Culling programs, by wiping out genetic lines in a relatively small group of animals, may significantly reduce genetic variation, producing, ultimately, some of the harmful effects caused by inbreeding.

If you are among the thousands of people who have followed elephant management in recent years, you know that chaos has already been visited upon the African elephant, brought by poachers who kill the animals for their tusks. In large parts of Africa, culling elephants is not the question. Poachers have already killed off so many elephants that it is not clear whether the creature can survive. The few that do survive in those areas are youngsters with no adults to guide them. The old bulls are gone. As Oria Douglas-Hamilton, Iain's wife, says, the social fabric of elephant life has been torn asunder, and whether it can ever be mended is in doubt.

The Ivory Wars

For the past 30 years political and economic instability has been rife throughout much of Africa. To achieve some form of economic stability, both personal and national, African leaders searched for some form of hard currency. One of the things they settled on was ivory. Like gold, it could be horded as a hedge against inflation. It became especially important when the price per pound of ivory began to sky-rocket. Until the late 1960s, ivory consistently sold for about $2.50 a pound. Then the price began to creep upward, cent by cent, then dollar by dollar as demand increased. Why the increased demand? Perhaps people in North America, Europe, and Japan simply began to have too much disposable income, money to be spent on beautiful luxury items. In any event, these nations created the demand, Africa provided the supply, and the elephant paid the price.

By 1973, a pound of ivory brought $14. Five years later the price had nearly tripled. In the 1980s it rose as high as $114 a pound in Oriental markets, where most of the raw ivory was worked. Because trade in ivory was illegal in many of the nations hardest hit by poachers, little or none of that money helped solve African economic problems. The poacher on the ground, who might be a barefoot villager armed with an archaic military rifle from World War I or who, in recent years, might carry a modern automatic weapon supplied by ivory dealers, might get $6 per pound. But even that was enough to make risking one's life worthwhile. In nations with some of the lowest per-capita earnings in the world, killing just two elephants could win a poacher what the average citizen would make as a year's wages.

Several factors helped spur the immense elephant slaughter of the 1980s. One was political instability. Guerilla revolutionaries and the government soldiers who pursued them through the wilderness often killed wildlife with complete indiscretion, often just for fun. Ivory could be used by governments and guerillas alike to fill war chests. African arms imports increased tenfold in the 1970s, ostensibly as part of a policy designed to quash political unrest. In eastern Africa, armed

In this undated photograph, African natives hold 6- to 8-foot-long elephant tusks, which will be shipped to an ivory-working port in the Far East. Improved trade routes and increased demand for ivory piano keys, billiard balls, carvings, and other products combined to pose a serious threat to elephant survival toward the end of the 1800s.

forces trebled. Historically, whenever arms and armies have increased in Africa, wildlife has declined.

Meanwhile, in some regions elephant protection instituted in recent years had helped the animals rebound from the wholesale massacres by ivory hunters early in the century. Kenya, for example, banned hunting in the 1970s, and several other nations soon followed suit. But while elephants were increasing, so were local peoples. As croplands spread into wilderness, people and elephants became neighbors. The ease with which elephants could be located and killed doubtless helped turn many a farmer into a poacher.

In 1980 Iain Douglas-Hamilton completed a survey of elephant populations in Africa and concluded that about 1.5 million elephants still roamed there. But poachers were going about their work so efficiently that his data were no sooner out than they were obsolete. Today, perhaps 600,000 survive, perhaps fewer than half a million.

Numbers alone do not tell the story. Poachers go after the elephants with the biggest tusks, so bulls are killed off first. When they are gone,

the guns are turned upon cows and even calves with rudimentary tusks. As the tusks collected become smaller, more and more elephants have to be killed to meet the undiminished demands of the market. Hong Kong, a leading ivory market, imported 521 tons of ivory in 1979, representing the deaths of 31,000 elephants. Nine years later only 290 tons were imported, but because average tusk size had shrunk by about half—to only 10 pounds each—that much smaller import still accounted for at least 33,000 elephants.

The slaughter ravaged elephant populations throughout Africa. Kenya lost 85 percent of its elephants, Uganda 90 percent. Tanzania lost 22,000 elephants a year during the 1980s. Naftali S. Gwae, a law enforcement officer in Tanzania, said that the 3,000 elephants that roamed Serengeti National Park in the 1980s were reduced to 450 within six years. Only half of the 109,000 elephants that lived in Tanzania's Selous Game Reserve in 1977 survived the next decade. Chad's 15,000 elephants were cut to 2,000, and the elephants of southern Sudan were nearly wiped out. An aerial survey of Somalia in the late 1980s tallied more dead than living elephants. The 90,000 elephants that lived in the Central African Republic in 1976 were cut to only 15,000 nine years later. The Republic, inhabited by both bush and forest elephants, exported nearly 2.5 million tons of ivory during the 1980s.

As the killing emptied out the open bush country, where elephants were easy to find, poaching spread to the rainforests. In Cameroon,

▶

A nervously smiling official shows the size of some of the tusks confiscated from poachers in Tsavo East National Park, Kenya. The country's elephant population was devastated by poaching and drought in the 1970s and 1980s.

Pygmies gave up a centuries-old tradition—the hunting of elephants with bows and poisoned arrows—in exchange for modern rifles provided by ivory smugglers. Modern weapons combined with the Pygmies' formidable hunting and tracking skills created a dangerous new force aligned against that forest elephant. In the 1980s, forest poaching increased in the Congo, too. As roads are cut into the protective forest, making way for oil development and settlement, elephant poaching there will probably escalate further.

In Zaire poachers took down about 200 animals daily through the 1970s. Data from Zaire is scarce, but experts believe the killing has not slowed. Some evidence suggests that Zaire's government itself is involved in the illegal ivory trade. In 1991 a National Audubon Society documentary film crew visited the Central African Republic and found that officials in one town, including the mayor, were heavily involved in poaching. The elephants of the Republic are under assault from all sides. Muslim traders and government officials supply rifles to the Republic's forest Pygmies, and Sudanese poachers with camel caravans have penetrated the forests. These poachers kill not just elephants but many other protected species, including leopards and hip-

▶

pos. In this nation, the size of Texas, the Ministry of Water and Forests does not even have one airplane to patrol against poachers.

The blame for the slaughter ultimately must be placed at the feet of those who purchase the trinkets into which ivory is made. These are entirely luxury items for which nonivory substitutes are readily available, and almost all the items are sold in Japan, the United States, and Europe. If you want to see what drove the killing of the African elephant, walk down New York City's Fifth Avenue and look at the ivory carvings in the shops there, or wander among the shops of Chinatown in New York City, San Francisco, or Los Angeles. The carvings and their price tags are there. Look too at shops in London, Antwerp, and Paris. The carvings you will see are the rubble of the elephant's collapse.

When the slaughter was at its height, Japan bought roughly half of all the ivory that Africa had to offer, nearly 500 tons yearly through the mid-1980s. About a third of it was carved into *hankos,* small seals bearing an individual's personal hallmark. Dipped in ink, they are used to sign checks and official documents. During the 1980s, Japan produced about 20 million hankos. The Japanese refused to use hankos made from other materials because, they said, ivory carries ink better than wood or plastic.

About a third of Africa's ivory found its way into the United States and Europe. In the 1980s, U.S. trade in raw and worked ivory claimed an estimated 35,000 elephants yearly, according to figures compiled by the Traffic division of the World Wildlife Fund. U.S. leather industries accounted for another 7,000 elephants each year, with little or no overlap between animals killed for ivory and those killed for skins. Thus, each year during the 1980s, the remnant parts of 42,000 elephants reached the United States in the form of luxury products such as figurines, jewelry, and expensive footwear. Although nations that claimed a surplus of elephants culled them legally and sold their products under an export quota system that was supposed to limit carefully the number of animals killed, fully 80 percent, perhaps even 90 percent, of the ivory that reached the big markets was taken illegally from poached animals.

The slaughter has wreaked havoc among the surviving elephants of hard-hit nations in eastern and central Africa. In those areas hardly an elephant population, or even a single herd, went unscathed. Out of several hundred Lake Manyara elephants studied by Iain Douglas-Hamilton, virtually none over 30 years old survived the 1980s. The loss to science alone was tremendous, since these elephants were the longest studied in the world, the first to provide keys to unlocking the secrets of elephant behavior and ecology. The longevity of the study had promised to provide data that is still years away in other areas.

Throughout much of Africa, bulls have been particularly hard hit. All the old bulls in Kenya's Tsavo National Park—the individuals crucial to good breeding—were wiped out. The bulls that comprised half the elephant population in Tanzania's Mkomasi game preserve in the 1960s were all wiped out 20 years later—a 1988 survey found no adult bulls there. In areas with such low numbers of males, chances for successful breeding are slim.

Old-ivory carvings in a New York City shop attract buyers from around the world. The difficulties of proving the legal origin of an ivory carving make controlled trade impossible.

Mature females—the matriarchs particularly—have fared little better. Only 3 of the 125 matriarchs that lived around Lake Manyara when Douglas-Hamilton started his study in 1965 were still alive in 1989. Young females, many in their teens, are trying to lead herds of orphaned babies and juveniles, a role that the teenagers—but for the slaughter—would not have assumed for another 30 to 40 years. They lack the knowledge of terrain and of food and water supplies that the old females would have taught them and, observers say, lead their groups as if dazed and confused. Over much of Africa the old herd structure, with females in their forties and fifties leading family groups bred by bulls over 40, will not recover for another 20 or 30 years, and then only if the killing is stopped. It is not clear how the elephants will recover the essential knowledge carried in the memories of their lost elders, or whether they can. The demand of the market place for ivory has left a perhaps indelible impression upon the elephant's world. Had the killing gone on at the incredible rate of the mid-1980s, the elephant would have been wiped out over much of Africa. It would quite likely have been only a matter of time before the destruc-

tion became complete. Fortunately, trade in ivory almost completely evaporated.

The Ivory Trade Ban

Since 1973, under the Convention on International Trade in Endangered Species of Wild Fauna and Flora (CITES), nearly 100 signatory nations have sought to control trade in wildlife products by classifying commercial species in two categories, called appendices. No trade is permitted in species listed on Appendix I. Limited, carefully monitored trade is allowed in products made from species on Appendix II.

The African elephant has been an Appendix II species since the treaty was in effect, so each CITES member that produced ivory was supposed to set an annual export quota on tusks. Beginning in 1985, tusks were supposed to be stamped with registration numbers so they could be tracked. If the system had worked, and if tusks had been taken only from carefully managed herds, the production of ivory under CITES should not have jeopardized elephant populations.

Unfortunately, the CITES system is strictly a monitoring program. The CITES agreement confers no *enforcement* powers upon the officials assigned to monitor trade, and each nation follows CITES rules and regulations at its own initiative. Any nation may ignore a CITES rule simply by registering an exemption saying it will do so. In addition, quotas and registrations often are abused. For example, in 1988 Somalia declared that it would export 8,000 tusks, even though the nation had only 4,500 elephants that year, many of them young. Burundi was even more abusive, more or less setting a record for CITES abuse. From 1975 to 1988, Burundi produced more than 2,000 tons of ivory *even though the nation had no elephants.* Officials claimed that the ivory had been seized from smugglers, but Burundi was actually cooperating with poachers.

CITES officials themselves did little to hamper illegal trade in ivory. In 1985, the first year that CITES required tusk registration, Burundi was permitted to sign up 90 tons of clearly illegal ivory to induce the nation to join CITES. CITES officials said that taking this illegal ivory off the market would also make it easier to monitor trade in the future. But Burundi quickly stocked up another 90 tons of ivory and, as recently as the summer of 1989—when the full effect of the poachers' work was well known—was permitted by CITES officials to register and sell 28 tons. This was a bonus to Burundi because illegal ivory sells for about $25 a pound, at most a fourth the price of legal tusks.

Singapore also benefited from the CITES legalization program. In 1986, Singapore's ivory imports reached a record 300 tons, in part because ivory merchants knew that CITES would legalize poached ivory that year to clear the market. When CITES did just that for Singapore in November, the value of the 300 tons increased literally overnight by millions of dollars. Ironically, and sadly, much of the ivory was stockpiled specifically to take advantage of the CITES program. Designed to make it easier to control poaching, the legalization program seems only to have intensified poaching.

◀

An ivory worker in Osaka, Japan, prepares a tusk for carving at the country's largest manufacturer of hankos, personalized seals used as signatures. In the 1980s, Japan bought about half of all African ivory on the market. Much of that worked ivory found its way into American and European shops.

Ivory travels from Africa to a variety of ports. In the 1980s the biggest entrepôts were Hong Kong, Singapore, Taiwan, Japan, and the United Arab Emirates. In the United Arab Emirates, carving factories made full use of a loophole in CITES regulations that limit trade restrictions only to raw ivory while automatically treating any worked ivory as legal. By definition, worked ivory is any tusk that has been cut into pieces or even mounted whole on a wooden board. Emirate carving factories were thus able to "legalize" thousands of tons of poached *raw* ivory by crudely cutting it so that it would qualify as *worked* ivory. The laundered ivory was then sent to Singapore, Hong Kong, and Japan for final carving.

Massive amounts of ivory moved through the markets in the 1980s—more than 1,000 tons yearly, compared to 200 tons in the 1950s, when average tusk size was larger. Nothing else showed so clearly that the CITES ivory-quota system was failing, that something else had to be done, and done soon.

At that point, with international cooperation a dismal failure, individual nations stepped in with protective plans of their own. In September 1988 the U.S. Congress enacted the African Elephant Conservation Act, authorizing the Secretary of the Interior to prohibit the import of ivory from any nation dealing in illegal ivory or failing to cooperate with CITES. As a result, ivory imports from Chad, Ethiopia, Gabon, and Somalia were outlawed. The law also banned the import of ivory from countries that act as intermediaries for nations dealing in illegal tusks. The following June, the United States—which had imported $18 million to $26 million worth of worked ivory yearly in the 1980s—announced a total ban on commercial ivory imports. Almost simultaneously, Japan banned the import of ivory from nations that do not produce their own ivory. The following September, Japan tightened the screws by banning all ivory imports regardless of the nation of origin. The Japanese ban even prohibited the import of elephant trophies by sport hunters.

African nations hard hit by poaching did not believe that such nation-by-nation bans were enough. Poachers were still killing elephants at about 20 times the rate that the elephant population could sustain. Consequently, Tanzania, at a CITES meeting in October 1989, asked that the African elephant be moved to Appendix I, which would ban all trade in ivory. Kenya sided with Tanzania, as did either other African nations.

Several nations of southern Africa—Zimbabwe, South Africa, Botswana, Malawi, and Namibia—opposed the ban because, they said, their elephant populations were stable or increasing. They wanted to continue to cull, which they considered necessary to sound management of the elephant. It was well known that these nations sold the tusks of culled elephants, though the primary reason given for the culling was the need to keep elephants from overpopulating.

Ban proponents argued that only a complete ban on the ivory trade could save the elephant. They said saving the elephant was critically important to economic survival: African nations garner little money from the ivory trade—less than 10 percent of the $60 million in legal trade reached national coffers—but elephant-related tourism in Kenya

A carved statue and tusk at the U.S. Fish and Wildlife Service office in Baltimore, Maryland, are among ivory pieces confiscated from travelers and importers. U.S. trade in ivory products helped kill some 35,000 elephants a year in the 1980s.

alone is worth about $200 million, more than 30 times the continent's entire trade in ivory. If the southern nations had too many elephants, ban proponents said, they could still cull them, ban or not. They simply could not sell the tusks from culled animals. After all, the southern governments had always maintained that culling was a biological necessity, not a business for profit.

The proponents won, though Zimbabwe, Botswana, South Africa, and Mozambique announced intentions to ignore the ban, and Burundi announced that it wanted to sell nearly 90 tons of ivory. There was also widespread consternation and disappointment when Great Britain allowed its protectorate, Hong Kong, six months to sell several hundred tons of stockpiled ivory despite the ban.

Nevertheless, the ban, coupled with increasingly tough control measures in Kenya and Tanzania, had tremendous results within months. In Somalia the price for a pound of ivory dropped from $35 to as little as $1, with no takers. The Central African Republic saw prices fall by as much as 70 percent within six months. In Zaire prices fell 60 to 85 percent. In South Africa the government that sold raw ivory for $150 a pound in January 1989 did not get a single bid 10 months later when it offered to sell 1.3 tons. Namibia and Yemen also found themselves without buyers, and lack of buyer interest forced Zimbabwe, Botswana, and South Africa to cancel an auction.

In some nations, a collapse in the ivory market predated the ban.

In May 1990 the World Wildlife Fund surveyed the 15 largest whole-sale ivory dealers in the United States and learned that ivory sales had begun to fall off months before the official ban. After the ban, consumer demand virtually dried up. Prices were down 40 to 70 percent for ivory jewelry and simple carvings, items that usually account for the bulk of the ivory business. The sale of high-quality carvings fell 5 to 10 percent. Hawaii, the largest U.S. wholesale market, saw an all-but-complete collapse in ivory sales. Demand, said one New York dealer, no longer exists.

In China, Japan, Hong Kong, and Taiwan the ivory markets soon collapsed after a jump in prices right after the ban took effect. Ivory workers were laid off in large numbers, and government programs were established to train them for other work. In China, failing ivory factories paid workers 75 percent of their salaries *not* to carve ivory.

The collapse of the market, brought about by consumer refusal to buy ivory, led to a nosedive in elephant poaching in Africa. The killing ground nearly to a halt in Kenya and Tanzania. Although stepped-up law-enforcement efforts in the African nations contributed to the decline in poaching, the killing would have gone on apace had not consumers refused to buy ivory, a commitment reinforced by the ivory-trade ban.

And What About Tomorrow?

Poaching is not the only obstacle to elephant survival. Another critical problem, and one more difficult to solve, is encroachment by people on elephant habitat. The elephant needs large quantities of food and water, and as human farms and settlements press against the boundaries of parks and reserves, the elephant is being squeezed into smaller

Kenya Wildlife Service director Richard E. Leakey addresses an American audience in Washington, D.C., in April 1991, stressing the need to continue the ivory ban. Leakey has been instrumental in making Kenya a leader in the fight to save the African elephant.

units of habitat. Whether it can survive locked into reserves, with no room for migrations during droughts and other times of food failure, is the knottiest problem faced by the Africa elephant and the people committed to its protection and management.

The African Elephant Conservation Coordinating Group, a panel of biologists formed in 1988 to help ensure elephant recovery and funded primarily by the World Wildlife Fund–U.S., the U.S. Fish and Wildlife Service, and the European Economic Community, has developed a plan for elephant management. It has targeted 49 elephant populations, ranging from 150 to 29,000 animals each, whose protection the committee believes is essential to elephant survival. The total number is less than 220,000, roughly a third of the elephants alive today in the more or less post-poaching world.

Richard Leakey, head of Kenya's Wildlife Service, believes that the coordinating group has set its sights too low. He says that existing parks can support some 700,000 elephants. But protection of even that low number will require intensive conservation measures. As outlined by the coordinating group, these include stepped-up anti-poaching efforts, better controls on trade in elephant products, development of management plans for each elephant population, and education of the public—particularly of peoples living within elephant range—about the elephant's plight and about the human role in creating that plight. Most important, because it is instrumental to the creation of sound management plans, funding must be provided for extensive research on elephant behavior, including the use of satellites and radio-tracking equipment to determine elephant movement patterns and habitat use.

The cooperation of Europe, the United States, and Japan will be critical to the success of the plan. These nations sustained the poachers and need now to make amends. They can do so by funding African elephant conservation programs. Without such outside help, the African nations, faced with the fastest-growing human populations in the world, will not be able to protect the elephant. As populations and poverty increase as if moving in lockstep, Africa may feel forced to open its parks and reserves to human settlement, offering temporary relief at best to an unmanageable human problem.

We are trying to help out—if in limited ways. Under the African Elephant Conservation Act, Congress authorized $5 million yearly for elephant protection. Since then, Congress has been slow to provide. In 1989, only $500,000—a tenth the authorized amount, about one thirty-second of the value of the ivory sold in the United States every year in the 1980s—was appropriated. Private citizens were more generous, funneling money to Africa through organizations such as the World Wildlife Fund, which spent nearly $3 million on elephant conservation in nine African nations in 1990. This money is needed not just to stop poachers but to fund the research needed to achieve the coordinating group's plan.

That a beast with the size and power of the African elephant, roaming one of the largest, wildest continents of the world, should be falling before the onslaught of humanity shows clearly the extent to which the fate of all wildlife lies in our hands. Creatures that only a

A cattle egret gets a free lunch and a lift from a savanna elephant in Kenya. The egrets feed on a trail of bugs stirred up behind the pachyderms. Researchers use unique ear markings to help identify individual elephants.

◄

few years ago loomed in the popular imagination as indomitable, even dangerous, have been rendered acutely vulnerable.

If civilization and culture are created within each of us by the process of learning to restrain one's destructive potentials—which is what most of our laws and mores are designed to do—then our success in protecting creatures such as the African elephant from our own capacity to overpower the Earth is a measure of the level to which humanity has risen. Similarly, our destruction of the elephant and of other species is a measure of how heedlessly we continue to conduct our lives, taking full advantage of our power to destroy one of the irreplaceable wonders of the natural world.

As Romain Gary asked in his novel about elephants, *The Roots of Heaven,* if there is no room in the world for the elephant, if we cannot help destroying this creature that asks mainly to be left alone, then how will we ever stop destroying one another? Where will we draw the boundary within which we will contain our ability to destroy wild lands and wild creatures?

The answer may lie somewhere in Africa, in those places visited centuries ago by single-masted ships, where elephants once moved in great herds across vast marshes but now are gone forever.

ROADS LESS TRAVELED

I n the late 1970s the world's newspapers, magazines, and television screens came alive with the gory tale of the Gulf of St. Lawrence seal hunt. The hunt took place each March on floating ice fields west of Magdalen Island, when harp seals inched onto the ice to give birth to pure white babies. Canadian hide hunters would be on hand every year to greet the newborn babies with bludgeons, clubbing thousands of them to death and flaying them of their snowy coats. The hides were then sold, primarily in Europe, to be made into trendy garments.

When the story of the annual slaughter hit the news media, people around the world were horrified. Demand grew for a ban on the killing of baby seals, although the sealers suggested that a ban would be enforced only over their dead bodies. Battle lines were drawn. Animal rights activists went to Magdalen and spray-painted the white baby seals to render their hides worthless. Confrontations between impassioned antisealers and equally impassioned prosealers led to fisticuffs on the killing grounds.

Despite the sealers' stalwart defense of their craft, time was running out for sealing. In a relatively short time the European Economic Community banned import of baby-seal hides, and the industry started to collapse. Finally, in 1987, the Canadian government prohibited the killing.

But that was not the end of the story.

Throughout the entire baby-seal wars, the seal hunters were incensed, angry, and anxious. They did not like outsiders intruding on their business, and they particularly did not like the idea of losing their income. If the sealers had only had a good crystal ball they might have divined that the intrusion had only begun and that the loss of income was, at very worst, only temporary.

The wars had brought the baby seals to the attention of millions of people worldwide, and a great many of those people, now that they knew the baby seals were there, wanted to see them. The result: Three years after the Canadian ban went into effect, baby seals were once again big business on the ice floes west of Magdalen Island in the Gulf of St. Lawrence. Income from tourists yearning to see live baby seals *tripled* the money once generated by baby-seal hides. And showing outsiders around the ice probably isn't any harder than whacking baby animals over the head and skinning them. It may even be easier, but at three times the income, no one really cares.

The travelers have probably helped ensure that the sealers take a new, more protective view of baby seals and of natural resources in general. In a sense, what has happened in the Gulf of St. Lawrence is a microcosm of what is happening in many natural areas throughout the globe. On every continent, and on quite a few of the islands in between, it is becoming clear that people can make a much better living by protecting the environment and showing it to tourists than

◄ G reen cornlilly and wildflower, Glacier National Park, Montana.

they can by destroying the environment and taking whatever short-term profit they can make.

The whole idea is predicated on a new type of tourist. This is the *ecotourist,* one who engages in ecotourism, which is defined as purposeful travel to natural areas to understand the cultural and natural history of the environment while taking care not to alter the integrity of the ecosystem and, at the same time, providing economic opportunities that make the conservation of natural resources financially beneficial to the inhabitants of the host region. In short, when ecotourists visit a wilderness, they leave behind only footprints and money.

Tourism is the second-largest industry in the world. The figures that have given tourism that position are little short of incredible. For example, each year nearly 400 million international tourists—mostly from North America, Europe, and Japan—spend about $55 billion dollars in developing nations. Worldwide, tourists spend nearly $200 billion yearly in the course of domestic and foreign travel, creating some 75 million jobs. That comes to a full 7 percent of the world trade in goods and services, and 33 percent of the same trade in developing nations. According to the World Tourism Organization, about 10 percent of the tourism market is "adventure travel," of which ecotourism is a component. About half of all special-interest travel by U.S. citizens is ecotourism.

U.S. ecotourists tend to be middle-aged, professional or retired, and relatively wealthy. A survey of U.S. tourists in Ecua-

---

ECOTOURISM

THE ISSUE Ecotourism, which is tourism conducted primarily to see natural places and things, offers promise as a means for protecting wild places by making them economically important. On the other hand, ecotourism can damage wild areas by bringing in more tourists than a region or its wildlife can tolerate.

THE CAUSE The desire to see exotic wilderness areas brings large numbers of tourists to places such as Africa, the Galapagos Islands, and the Latin American rainforests. Money spent by these people gives governments in poorer developing nations a reason to protect wildlife areas rather than develop them for only transient economic gains. However, as areas become popular, they tend to bring in more people than local governments and businesses can manage. This may have harmful effects on wildlife and wildlife habitat.

EFFECTS ON WILDLIFE Tourists can destroy habitat or disrupt necessary wildlife behavior patterns, such as foraging and breeding. However, tourists can also protect wildlife and its habitat by providing funds for wildlife management and by discouraging illegal activities such as poaching.

EFFECTS ON HUMANS Ecotourism can provide needed income to people in developing nations, creating jobs and outlets for goods. It can also generate funds that support both local governments and the national economies. For example, ecotourism is Kenya's largest source of foreign income. On the downside, an influx of tourists can damage local cultures. Also, if local people are not trained for management positions, ecotourism can lock them into dead-end service jobs.

EFFECTS ON HABITAT Ecotourism can stimulate the protection of wild places, but is can also lead to habitat damage through overuse. In parts of Africa, for example, use of off-road vehicles is damaging grasslands and adversely affecting animal behavior.

The National Audubon Society Travel Ethics Code

1. Wildlife and wildlife habitats must not be disturbed.
2. Tourism to natural areas will be conducted in a sustainable manner.
3. Waste disposal must have neither environmental nor aesthetic impacts.
4. A tourist's travel experience must enrich his or her appreciation of nature, conservation, and the environment.
5. Ecotourism must strengthen the conservation effort of the nation visited and enhance the natural integrity of the places visited.
6. Engage in no trade in wildlife products that threatens plant and animal populations.
7. The sensibilities of other cultures must be respected.

dor indicated that more than half had a family income of more than $30,000, and a quarter earned more than $90,000. They also spend relatively freely. A World Wildlife Fund study by Elizabeth Boo established that people who visited Latin America primarily to see national parks spent $1,000 more per capita during a two-week trip than did other tourists.

Even in the United States, ecotourism is a booming business. Recreational visits to the national park system alone are approaching 300 million yearly. Tourism in Wyoming, home of Yellowstone National Park, brings in nearly $1 billion each year from wildlife-related activities. Yellowstone alone has more than 2.5 million visitors a year, which is roughly five tourists for every Wyoming resident.

Tourism is a growing business, and although no solid data exist for measuring it, ecotourism, according to tour operators, is a rapidly growing part of the larger industry.

This offers a lot of promise to some of the Earth's last wild places. From volcanoes to coral reefs, from Rocky Mountains to African plains, tourism is providing an economic justification for the protection of wildlife and its habitat. But ecotourism is a double-edged sword. As the number of tourists visiting any one wild place grows, ecotourism threatens to turn abruptly into ecoterrorism, the sheer mass of visitors trampling wilderness into mush.

A visitor to Yellowstone National Park watches the sun set among the geysers. Sometimes, for a moment in the world's great parks, visitors feel at one with nature.

The Protective Promise of Ecotourism

Of the $55 billion that tourism brings to the developing world, as much as $12 billion may come from ecotourism. For individual nations, the impact of ecotourism may be quite substantial. In

A vacationer in Montana's Glacier National Park finds solitude while floating on Swiftcurrent Lake. The historic Many Glacier Lodge gives the scene a hint of the Swiss Alps. The park offers some of the most inspiring natural experiences to be had in this country.

Passengers of mid-1930s rolltop touring cars hop out to enjoy a view of Jackson Glacier in Glacier National Park. The cars are used to ferry visitors around the park's attractions.

A lioness at Nairobi National Park, Kenya, seeks shade from the midday sun. Some 650,000 tourists come to Kenya each year to see big cats, elephants, and rhinos, and to spend half a billion dollars.

Rwanda, Kenya, Costa Rica, Ecuador, and Nepal, ecotourism is a leading generator of foreign revenue. The development of ecotourism as a major source of revenue is helping to protect wildlife in many of these nations because the national economies need the income generated by wildlife-related tourism.

Kenya is a classic example. Every year some 650,000 visitors pass through Kenya's national parks and protected areas, spending nearly half a billion dollars. That income has made tourism the nation's leading source of foreign revenue since 1988, when tourism overtook coffee and tea as Kenya's top money earner. About 17,000 square miles of Kenya—fully 8 percent of the nation—is protected in 52 national parks and reserves. Forest reserves account for another 3 percent.

Surveys show that most tourists come to Kenya to see elephants and rhinos. Both creatures have been hard hit by poachers, rhinos almost to the point of extinction within Kenya. Elephants have fared little better. Some 70 percent have been lost on protected lands and more than 90 percent in unprotected areas. But Kenya has been among the leaders in putting an end to poaching, and since an international ban on ivory was instituted in 1990, the elephant decline has leveled off. Rhinos also have benefited from increased protection and better training of game rangers.

Protecting the animals has little to do with altruism. The average elephant herd is worth an estimated $610,000 per year in tourist income, and an individual elephant is thought to be worth close to $1 million during its lifetime of some 60 years. The World Bank helped Kenya develop Amboseli National Park as a tourist attraction in the

1970s because the return from agricultural use of the land could be expected to accrue only 33 cents an acre, but tourism would bring in nearly $17 an acre.

In an especially bold move, Kenyan president Daniel arap Moi established the Kenyan Wildlife Service as an autonomous agency administered by a private board of trustees. Under this new system, which emphasizes protection of income-generating wildlife areas, all revenue raised by the national parks and reserves remains under the jurisdiction of the wildlife service, which allows the money to be invested in better conservation of the protected areas. The park system is run as an independent corporation that must balance its income and expenditures.

One critical problem for any protected area in a developing nation is the interest of local people. Historically, local residents have been ignored when protected areas were established. In some cases, protection actually meant a loss to native peoples, since hunting lands and potential crop fields were taken from them. The problem was compounded when wild animals roamed freely from the protected lands and damaged crops or competed with domestic animals for water. This created strong resentment among local people, who felt they were losing money to protect lands and animals for foreign visitors, and helped encourage poaching. Landowners put up fences to keep wildlife away from scarce water sources and to stop movement out of the parks. Consequently, wildlife declined in many parks.

Recognizing the value of wildlife to the burgeoning tourist industry, the Kenyan government has moved to assuage the concerns of local residents. One method has been the provision of financial incentives. For example, the government requires that tourist lodges must preferentially employ Kenyans in all but the most senior positions. Lodges must use Kenyan food products as much as possible and keep the serving of imported foods to a minimum. Every game-lodge visitor pays a $5 tax that is put into a trust fund for nearby private landowners. Parks that have no lodge deposit a percentage of gate receipts into local trusts. The money is subsequently used to fund schools, hospitals, and the treatment of cattle for insect pests. Remaining revenues are distributed to individual landowners.

The incentive system is young, but it shows signs of working. For example, when the Maasai Mara County Council Game Reserve was established in 1960, it severely hampered the movements of the pastoral Maasai people and put their cattle into direct competition with grazing wildlife. The Maasai's resentment of the park led to serious poaching, even though the Maasai traditionally are not a hunting people. Now the government and the Maasai share park revenue. Every visitor who stays at one of the 11 lodges located in and around the reserve pays $10 as a sort of surtax. Half of it goes to the county council that administers the reserve, and half into a trust fund for local Maasai. The Maasai now earn more than $1 million yearly. Because wildlife is an asset to them, poaching has virtually stopped. As one Kenyan axiom puts it, "Wildlife pays, so wildlife stays." Poaching in other parks where local residents have been similarly accommodated also has declined.

A classic example of how wildlife can benefit from ecotourism is

A large bull forces tourists to back off and let a herd cross the road in Amboseli National Park, Kenya. One elephant represents hundreds of thousands of dollars worth of foreign currency brought into the country through tourism. ▶

A Maasai woman repairs her roof, with the Maasai Mara Game Reserve in the background. Native tribes are earning more than a million dollars a year through revenue-sharing programs. ▶

Rwanda's Parc National des Volcans. Lying among the forested Varunga Volcanoes, an area made famous by the work of biologist Dian Fossey, the park is home to what was for many years an ebbing population of rare mountain gorillas. Poachers were slowly killing off the animals to sell skulls and hands as tourist souvenirs. The government of this tiny central African nation could not afford to protect the gorillas adequately, and it seemed likely in the early 1980s that poaching would wipe out the animals, aided by the cutting of the gorilla's forest habitat to make way for crops and cattle.

The solution to the problem proved to be tourists. By opening the area to ecovisitors, and charging each one $170 an hour to see the gorillas, the government succeeded in generating $1 million yearly in gate receipts. A mere $150,000 of that is paid in salaries for 70 park guards and up to 10 guides. The presence of the guards and, presumably, of frequent tour groups has reduced poaching to zero since 1983.

Ecotourism is also proving a benefit to the world's beleaguered tropical rainforests. For example, tourism in Manaus, Brazil, on the Amazon River, may ultimately help stave off rainforest destruction there. The number of tourists coming to see the forest jumped from 12,000 in 1983 to 70,000 five years later. By the close of the century, tourism is expected to explode to perhaps 200,000 visitors yearly. Those visitors come only if there is forest to see. They are not seeking burned-out lands or the snarl of chain saws. Similarly, the Caribbean island of Dominica, still covered by 50 percent of its original forest, is trying to protect its natural attractions to build up the tourist industry. Thailand, in Southeast Asia, is also trying to protect natural resources to make the nation attractive not only to tourists but also to American retirees seeking a dwelling place more exotic and less developed than Florida.

Costa Rica is attempting both to protect its rainforests and to develop them for tourists. In 1989, 375,000 tourists came to Costa Rica, according to Tanya D'Ambrosia, subdirector of commercialization for Costa Rica's Board of Tourism. About half of these people engaged in ecotourist activities, such as visiting a park. In 1988, tourism brought in $170 million, including $5 million from cruise ships. A year later, the total jumped to $206 million. Tourism has outstripped cattle as a source of income and is third in importance to coffee and bananas. As tourism dollars grow, so does the importance of preserving the rainforests.

But as tourism increases so does the need for proper protection and management of the places tourists come to see, notably the rainforest reserves and parks. Luis Manuel Chacon, the head of Costa Rica's Board of Tourism, likens the parks to a well. If you pump it with moderation, you will leave a heritage for your children. Pump it too heavily, and eventually you will have only mud.

Chacon has asked the Austrian government to help fund a study on the capacity of Costa Rica's parks and reserves to absorb tourists. He also wants to make local residents an instrumental part of park management. "They'll be the worst enemy of the parks if they can't make a living off them," he says. His model is the private Monteverde Cloud Forest Reserve, run by the Tropical Science Center, a not-for-profit group of tropical biologists. The land was donated by Quaker farmers

A tour group traipses through a St. Lucian (Caribbean) rainforest with a forestry department naturalist. All visitors to the island's rainforest must have a paid guide.

who have lived in the area since the 1950s. They donated it to help create a reserve that would not only protect the mountain forest but also preserve the watershed that feeds into their farms.

The reserve covers 24,000 acres and is increasingly popular with tourists. In 1973–74 the reserve had 403 visitors. In 1987 it had nearly 13,000. That year the reserve made close to $35,000 in lodging, entrance fees, and souvenir sales. The reserve has, says Chacon, developed a sort of ecotourism high school that teaches local people the essential knowledge needed to deal with foreign tourists, everything from basic biology to such practical tasks as how to set a table. The park also benefits local people by selling local crafts. One cooperative earned $13,700 for the 66 housewives who belong to it. High-school students in another program earned about $110 each in 1987 by coloring postcards. Though the total sums per person are low, they are substantial as supplemental income. Some students earn more per hour coloring postcards than their parents do working at other jobs. Since students have to pay relatively expensive school fees after the sixth grade, the money may mean the difference between continuing an education or dropping out.

Chacon wants to see rainforest preserves draw in more visitors, but he wants to keep development out of the parks. He would like to establish development zones around the parks, limiting lodges and restaurants to these zones. Visitors would then hike into the parks. He wants to establish a code that will limit all buildings surrounding parks to a height no greater than that of the tallest palm tree.

Many of Chacon's ideas about park management come from visiting parks outside Costa Rica in such places as Mexico, Spain, and the United States. He says that he tries to learn from the mistakes he has

observed in those nations. In the United States, for example, lodges and other tourist attractions are built right in the parks and, though they help to draw big crowds, they pose serious problems for sound park management.

The United States provides perfect examples of the dark side of eco-tourism. Many of the nation's protected lands are overrun with tourists. Yellowstone's 2 million acres swarm each year with the 2.5 million tourists who show up primarily in June, July, and August. Exhaust fumes from motor vehicles create serious air pollution problems along heavily traveled roads, and some of the nation's worst traffic jams occur in summer in towns surrounding the park.

Glacier National Park brings in about 1.5 million tourists each year, but few ever venture more than 45 minutes away from their pollution-generating motor vehicles. Glacier is a by-product, in a sense, of tourism. Congressional allocation of land for the park in 1910 was won primarily by railroad magnate Louis Hill, who supported the park as a destination for vacationers traveling on his Great Northern Railroad. Once the park was approved, Hill moved quickly to build a chain of lodges in the park to accommodate visitors. In those days, the park was visited mainly by the wealthy, who could afford long vacations and train travel. But that changed after World War I, as it would change for many of the nation's parks. The automobile came into popular use, and park visitation soared. At Yosemite National Park in northern California, trains brought in 14,251 visitors in 1916, but that was the year automobiles overtook trains as the visitors' preferred mode of transportation. Two years later, fewer than 4,000 came in on trains, while automobile travelers leaped to 26,700. After the war, motor vehicles made travel a national pastime. Today, Yosemite, Yellow-

A Yellowstone National Park bison-jam blocks traffic and causes nightmares for park rangers. More people are injured each year by bison and other seemingly harmless animals than by grizzly bears.

stone, Glacier, and the other parks—created as much to sell railroad tickets as to preserve local scenery—each draw millions of visitors yearly.

These millions of visitors to U.S. protected lands—nearly 300 million a year to the national parks alone—can barely be accommodated. They need lodging, they need health facilities—Yellowstone employs 40 emergency medical technicians and still witnesses nearly 20 deaths a year—they need restaurants, they need regulation and law enforcement, they need parking places, they need a place to put their garbage. Managing massive numbers of ecotourists is a difficult task even for a nation as wealthy as the United States. Increasingly, administrators are raising fees to reduce the number of tourists or are contemplating the need to restrict the number of citizens who can come each year into national parks, forests, and wildlife refuges. Thus, the industry that helped create many of these preserves and that continues to justify their existence is, at the same time, threatening to strangle the life out of them.

If that is a problem in a highly educated, highly developed, heavily taxed nation such as the United States, then what does growth in ecotourism portend in the developing world?

Ecotourism/Ecoterrorism: Dr. Jekyll and Mr. Hyde

Ecotourism has helped give nations in Africa, Asia, and Latin America sound economic reasons for protecting wildlife and wildlife habitat. But ecotourism also brings with it special threats and dangers.

A Columbian ground squirrel in Glacier National Park, Montana, conspicuously consumes tourists' potato chips, to the delight of its human audience perched above. Some visitors insist on sharing their junk food, despite signs pleading for restraint.

Both the positive and the negative aspects of ecotourism stem from a single source—the increasing number of people engaging in travel to wild places. The visitors whose amassed expenditures provide income to local residents and developing governments, thus creating a reason for protecting wild places, also, by their sheer numbers, threaten to destroy the areas to which they flock. In some of Africa's parks, for example, grassy plains are crisscrossed with the tracks of tourist vehicles. At Maasai Mara, tourists regularly give generous tips to drivers who take them off roads and across wild lands for a closer look at elephants, lions, cheetahs, and the like. Cheetahs in some Kenyan parks have been so thoroughly harassed that they fail to breed and often cannot hunt successfully. Herds of elephants and buffalo are often disturbed on their feeding grounds by tourists traveling by balloon. Forty-five minutes in a balloon costs each tourist about $250—a tidy income for the balloon owner—but when the balloons swoop low over herds, the disturbed animals waste precious feeding time and burn off calories.

Chimpanzees at Tanzania's Gombe Stream National Park are jeopardized by uncontrolled tourism. These are the chimps that biologist Jane Goodall has been studying since 1960. When she first came to Gombe, which was then a 30-square-mile reserve, the remote area was scarcely traveled by people. Now, for weeks on end, as many as a dozen tourists a day stop by hoping to see the famous champanzees. But the park has no facilities for visitors, who camp out on the shores of Lake Tanganyika and wander unguided through the forest. Marauding baboons have attacked visitors, and the chimps themselves pose a dangerous threat to people.

But the chimps, too, are in danger. In 1988, 13 of 48 chimps habituated to humans died of a flulike disease complicated by pneumonia, and six infant chimps were orphaned. The disease, Goodall believes, was probably contracted from tourists. Uncontrolled tourism and lack of sanitary facilities leave the chimps open to an epidemic that could wipe them out, says Geza Teleki, a Gombe researcher.

Another problem concerns the economics of ecotourism, which often do not work to the benefit of the wild areas visited by tourists. According to a report on ecotourism by Kreg Lindberg for the World Resources Institute, most governments set entrance fees for protected lands much lower than the price tourists would willingly pay. Because fees are low, they do not help to keep the number of visitors to a minimum. As a consequence, many parks have more visitors than they can carry without damage. Natural attractions are spoiled, and, to top it off, much of the revenue raised by the low entrance fees is put into a general fund rather than used to restore protected lands.

Yet another problem is the transient nature of ecoattractions. Generally, wrote Lindberg in his report,

◄

This young elk in Yellowstone National Park nearly strangled itself with a swingset. The animal positioned one swing under its neck and one under its stomach. In a panic, it catapulted itself about five feet into the air and fell to the ground on its neck, but then seemed to recover its composure.

> a particular region may initially attract a few "low impact" tourists who rely primarily on existing facilities. As word of the destination's appeal spreads and visitors flock to see it, however, residents begin to tailor facilities specifically for tourists. Heavily promoted facilities spring up, many of them foreign financed, and visitors become disenchanted. Major franchises dominate the supply of tourist attractions, eventually the region

stagnates, and revenues fall unless a deliberate effort is made to rejuvenate the region.

Foreign investment in lodges, tour-guide services, and other eco-tourism-related businesses in developing nations dilutes the benefits that ecotourism brings. The World Bank estimates that 55 cents of every dollar tourists spend in developing nations filters back to the developed world. For individual nations, the percentage is even worse. Zimbabwe loses 90 percent of every tourist dollar. These losses can be minimized by careful planning, for example—as in Kenya—by requiring foreign-owned lodges to hire local residents and to use local goods and foods.

Although larger numbers of tourists provide more money than smaller numbers, they also pose several problems. Most developing nations lack the revenue needed to manage heavy concentrations of tourists in remote areas. Because tourists are unregulated, they often damage the areas they come to visit. One solution to this problem is better education of tourists, but again, many developing nations lack the personnel and educational materials needed to instruct visitors on proper wilderness etiquette.

Ecotourists can affect the areas they visit in three adverse ways. They can cause ecological damage, rendering a protected area less attractive to future visitors. They can become so numerous that they get in each other's way, destroying the pleasure of the wilderness experience. And they can become so ubiquitous that local residents begin to resent them. Frequently, all three factors coincide.

A cheetah searches the plains of Kenya's Amboseli National Park for its next meal. The world's fastest sprinter is no match for thoughtless tourists. Cheetahs, being daylight hunters, are often interrupted by tour buses when hunting in national parks, reducing their luck at securing food.

All three problems can be minimized by limiting the number of visitors permitted in a protected area. But even in the United States, administrators are reluctant to take this approach. Nevertheless, other governments have set visitor quotas with some success. Rwanda permits only about 6,000 gorilla-seeking visitors into the Parc National des Volcans, but has kept profits at the 80-percent level—more than $1 million yearly—by setting high fees for gorilla watching.

Setting high fees is also a way that governments can attempt to limit access to protected areas. Park fees constitute only a small part of the cost of overseas travel, so even doubling them raises overall expenses insignificantly. And studies indicate that ecotourists are willing to pay more if they know their fees help protect the areas they visit.

The Galapagos Islands, 600 miles off Ecuador, are a perfect example of the Janus head that is ecotourism. The Galapagos apparently were unknown to humanity until 1535, when a ship carrying the bishop of Panama set sail for Peru, hit the doldrums, and was carried to the 3,000-square-mile cluster of 19 islands. The bishop left a written record of his visit—including descriptions of the marine iguanas and Galapagos tortoises, found nowhere else—but the islands did not appear on a known map for another 35 years. They were named the *Isolas de Galapagos*, the Tortoise Islands, after the islands' most distinguished natives. There were 14 known subspecies of tortoises on the island then, numbering perhaps 250,000 and weighing up to 600 pounds. For the next 300 years or so they were much valued by whalers and other seafarers, who captured them for food. The tortoises were great warehouses of meat—they could live for a year without food or water, and they did not move around very much. Today, only 15,000 remain, and three of the subspecies are extinct.

For centuries the islands drew the occasional scientific visitor. One was Charles Darwin, who spent five weeks there in 1835. His study of the island's finches was instrumental to his conception of biological evolution based upon natural selection. It was the interest and encouragement of scientists that helped stimulate the government of Ecuador, which claims the islands, to declare 97 percent of the island area a national park. The other 3 percent is used for military bases, agriculture, and urban development—the islands' 10,000 residents live in eight towns scattered on four islands. But despite the park designation, it was not until 1970 that organized tourism cruised into the Galapagos, with some 4,500 tourists arriving on ships that year.

Business has boomed. The government of Ecuador wanted to keep the number of tourists below 25,000 yearly, but in 1986 that number was topped by more than a thousand. By 1990 it had doubled.

Tourism is a boon for the islands and for Ecuador. Tourists bring in $700,000 in fees yearly, more than all the other national parks in Ecuador combined. The Galapagos Islands seemingly need all the funding they can get. The influx of tourists—which can reasonably be expected to grow rapidly within the next few years—is being felt like a nagging bellyache. Although all visitors are required to hire a professional guide, some of the islands' trails nevertheless are so badly eroded that they may have to be closed. A second airstrip was recently opened to accommodate burgeoning air traffic, and a third is in progress. The islands lack telephones and a hospital but now have crime

Tourists on Going-to-the-Sun Road delight in roadside attractions at Glacier National Park. Superintendent Gil Lusk wishes more visitors would leave the road and discover the park's real attractions, available only by foot or on horseback.

and prostitution. Persistent schools of human swimmers actually drove white-tipped sharks from traditional haunts along one beach until people were banned from swimming in the area. More hotels are being built, along with restaurants, dance halls, and souvenir shops. Jobless Ecuadorians hoping to make a killing in the tourism business continue to move to the Galapagos, increasing both the population and the various stresses on the land. Some natural sites are becoming overcrowded with tourists, and reports of illegal garbage dumping by boats are increasing.

But despite these threats, the outlook for the Galapagos is relatively good. The islands boast 70 licensed guides, each of whom has had a month of training. The guides restrict tourist activities to trails and permit no contact between visitors and wildlife. The Galapagos National Park Service works closely with the scientists of the Charles

Darwin Research Station on environmental education and wildlife conservation. The islands seem well supported at the national level: Ecuador's president in 1990 halted construction of a five-star hotel pending development of new tourism regulations.

If properly managed, most protected areas can easily accommodate large numbers of tourists. It is lack of sound management and control that turns ecotourists into ecoterrorists. Galapagos wildlife doubtless suffered greatly every time a few dozen whalers hauled ashore in the nineteenth century and started carting away tortoises, but today, 50,000 tourists can pass through without posing any danger, if they are well regulated. The same can be said for other ecotourist destinations. In the 1970s it was estimated that Kenya's Amboseli National Park could hold no more than 70,000 tourists, but today the estimate is 250,000, provided that the visitors and their movements are controlled. However, that involves educating tourists about the dangers they create when they drive off of roads or approach animals too closely, and many nations cannot afford the effort needed to train tourists to be good ecotourists.

Clearly, ecotourism offers great promise for the protection of wild areas. In today's world, to paraphrase a 1960s protest singer, money doesn't talk, it threatens, blackmails, and bludgeons, it generally has its way. This makes it a bad enemy to have against you, but a pretty

A fly-fisherman tries his luck in the icy Swiftcurrent Lake near Many Glacier Lodge in Glacier National Park, Montana. Despite a short summer season, Glacier is still relatively uncrowded compared to Yosemite and Yellowstone national parks.

►

The lower falls of Yellowstone National Park offer a feast for the eyes and ears. The park's many wonders attract more than 2 million visitors a year, a figure that increases annually.

A family engages in some good, old-fashioned stone skipping on St. Mary Lake in Glacier National Park. Though most visitors stay only a day or two and rarely venture far from their cars, they still come away with a sense of wildness and renewed appreciation for natural places.

good weapon to have on your side. Given today's social and political values as well as the demands of world hunger, the ultimate power of money suggests that wild places can be saved only by turning them into revenue generators. If we cannot do this, then the profiteers and money-makers of the world will find some other use for wild places, and they will be aided in doing so by the hunger and poverty of the developing world.

Ecotourism thus may play a crucial role in wildlife and habitat preservation worldwide. But it will succeed only if we ecotravelers carry with us a responsibility for the places we visit that goes beyond the mere payment of fees. Paying fees entitles us to visit protected areas only within the limits of sound wildlife conservation. It does not entitle us to travel roughshod wherever we wish in pursuit of whatever we wish to see. If developing nations cannot afford to educate us and monitor us, then we have to educate and monitor ourselves. We should know before we visit a tropical rainforest or an Africa savannah all that we need to know to ensure that our visit is a harmless pleasure.

If we assume our responsibilities as international travelers, we can reasonably demand that developing nations ensure that our dollars be funneled into better protection of wildlife and wildlife habitat. We can claim an interest and a share in the management of distant wildlife resources. And we can ensure that we do in fact leave behind only footprints and a fair share of our wealth and influence.

PLANS FOR PLANETS

 recent issue of *Life* magazine featured a cover story about the planet Mars. The story is called "Our Next Home, Mars: Bringing a Dead World to Life" (no pun intended, one presumes). The article describes Mars as it is today: air almost entirely carbon dioxide, deadly to human life; landscape barren and lifeless, with the possible exception of subterranean microbes or some long-dormant but still viable unicellular spores; nightly temperatures around minus 125 degrees Fahrenheit.

The story then gives us a fascinating account of how, through a process called *terraformation*, we will, during the next 200 years or so, turn Mars into a planet on which people can live. We will give it an atmosphere that can be breathed and that will retain solar heat, warming the planet. We will plant trees and other vegetation. We will bring in livestock. By 2150 more than a million of us will live on Mars, exploiting the planet's mineral wealth and taking advantage of a summer twice as long as Earth's to grow several crop harvests yearly. As more immigrants pile in over the next two decades, companies on Earth will take advantage of the cheap labor and make Mars a major production site. Mars will, says *Life* magazine, "be the richest colony in human history, the jewel in Earth's crown." Of course, the article suggests, living 35 million miles from home may give the new Martians a perspective on life that makes them rankle a bit at being treated, and exploited, as colonists. The article hints at the idea, but does not explicitly say, that the Martians may one day rebel and declare independence and so on. It also does not say, or even hint at the idea, that people on Earth—the Old Planet—may come to look down on those provincial Martians as being a bit backward socially, a bit unsophisticated, the way New Yorkers look down on midwesterners, and the way Europeans look down on New Yorkers. This tradition of the old and entrenched sniffing with superiority at the new and growing is an ancient one, and it is not likely to change in the next 200 years. A whole new caste system could arise.

The article does say, however, that the whole shindig should cost about $3 trillion, give or take a few billion. Which probably means that, given the vagaries of inflation and government spending, the cost will spiral into the quadrillions. But never mind, eventually those alive when the project is completed some 10 generations from now will begin to see a generous return on the investment because Mars will be an entire planet just there for the plucking. And you know—though the article did not mention this either—it could be a real entrepreneur's paradise, too, because those pesky environmentalists, who seem always to be suggesting that we should aspire to production without pollution and environmental destruction, probably will not have much to grieve over on Mars, no matter how bad things get. All you would have to do is look back to what Mars was like when it all started and you would have to admit that things could not get much worse. There apparently are not even any species to lose.

◀ **D**usk, White Sands National Monument, New Mexico.

Three trillion dollars. Of course, that amount would be spread out over a 200-year period that would begin just about *now.* In fact, it has already begun. President Bush, in one of his efforts to save vanishing federal agencies, has declared a "New Age of Exploration" as part of his recovery plan for an ailing National Aeronautics and Space Administration. The New Age of Exploration includes Project Mars, which is slated to land humans on Mars before 2020. Somehow, according to a nameless Bush administration spokesman quoted by *Life,* this "will power a revolution in science and technology, boost the economy, reanimate the educational system, and elevate national morale. On White House orders, blue-ribbon panels of aerospace experts have been developing a strategic plan to achieve the President's goal: 'The American flag should be planted on Mars.' "

For many NASA scientists this is the first stepping stone in terraforming Mars. As one biophysicist put it, "It's ridiculous to go all the way to Mars just to plant the flag, grab a few rocks and come home.

Industry around New York Harbor, as seen from an Environmental Protection Agency helicopter, helps make the area's water among the dirtiest in the world. Frequent oil and chemical spills, offshore dumping, and millions of automobiles add to the tab.

A mill worker herds logs toward the sawmill at a lumber operation in Lyons, Oregon. Our continuing removal of limited resources such as ancient forests has begun to catch up with us.

Humanity needs a new vision, a new challenge, not a cosmic park. Mars could provide that challenge.'' Well, of course! Who would want to go all that way just to plant a flag when they could, as long as they're there, terraform the whole place? And certainly we do need that new challenge, now that the old challenges back here on the Home Planet have become so dull, just commonplaces like world hunger and ozone depletion and grinding poverty and that old devil, global warming. Will those nagging old challenges never go away? Can we never get on with terraforming neighboring planets?

Well, breathes there the human with soul so dead that the idea of turning Mars into a living planet replete with people does not stir the blood and make the imagination crackle? You would have to be nearly brain-dead not to quicken at the idea of establishing an interplanetary civilization, of becoming a galactic species. But isn't it ironic that at the very time we are talking about turning Mars into something resembling present-day Earth, we are industriously turning Earth into something resembling present-day Mars? One astrophysicist has suggested we need to set up life on Mars so that we will not have all our eggs on one planet, just in case something really big goes wrong with Earth. But what do you think? Do you suppose there might be a simpler, more elegant solution to our earthly problems than trying to terraform Mars? Before we pay the first installment on the three trillion, do you suppose that perhaps we ought to take a close look at how we are terraforming our home base, and at what we can do about it? Do you suppose our president or the NASA scientists ever think about that possibility?

Terraforming the Earth

History, someone once suggested, is the story of a series of wars. You could also argue that it is the story of a series of culture-changing inventions. But in addition there is a hidden history of humanity, one largely ignored because the dangers in which we dabble are so veiled by our optimism that we fail to see them as hazards. This is the history of ecological change that human society has wrought upon the globe in the past 10,000 years or so.

Much of that history is embedded in prehistory, which helps to obscure the issue. We tend to forget that until roughly 15,000 years ago every human society consisted of hunter-gatherers, people who lived close to, and in harmony with, their habitat. Disease and the availability of forage took care of overpopulation. We did not seek to change our habitat because it was our source of sustenance and we valued it for what it was. Even more important, we did not know how to change it.

Some cultures learned. It started with agriculture. Planting crops removed a lot of the uncertainty that came from dependence on wild game and wild plants. The cultures that adopted agriculture as a mode of life freed themselves from certain biological constraints, at least for the moment. They forgot the old concerns that their forebears had about the integrity of natural habitat and wild places, concerns that survive today as a vague philosophy among some recently destroyed hunter-gatherer cultures such as those of the Native Americans. When early farmers moved out of the Middle East and into the lands of hunter-gatherer proto-Greeks and proto-Germans, the supplanted cultures doubtless looked on in dismay as their old hunting grounds were turned into crop fields. As some Native Americans do today, these ancient Europeans must have thought of how the circle of life had been broken, how the newcomers were turning the earth upside down, how the wild herds were vanishing. Then the people of the dying cultures were absorbed into the new. Soon they were farming too, cutting away the forests of Europe to create farmland. Centuries later their descendants—who had lost all memory of the old hunting times—discovered the New World, encountered existing hunter-gatherer societies, and repeated the cycle.

The history of human society is a history of habitat change. Since history began, in the form of angular markings on clay tablets, we have been working constantly at trying to change natural habitat into the habitat of agriculture, then of cities, and then of industry. We have labored with imagination and commitment. One has to be highly

Power company smokestacks on Boston's Mystic River stand as monuments to industrial strength and environmental neglect. The mid-1800s Industrial Revolution introduced an era of unsurpassed economic growth and ecological degradation.

imaginative to look at a forest and see croplands, waving seas of grain; to look at a marsh and see more cropland; to look at the naked shores of a protected coastal harbor and see the walls of a great port city. One has to be deeply committed in order to toil for years at the cutting and burning of forests and the draining of wetlands.

We were industrious and, some of us now think, reckless in our slavery to development. But who can blame us for our recklessness, which was founded on an optimistic belief that we were improving the world? All the signs and portents suggested we *were* improving the world. Agriculture fed more people than wilderness could. The Industrial Revolution made life so much easier. More people had more wealth, more people had more things, more stuff. In many nations people for the first time ever could aspire to rise above their initial lot in life. The human population grew by leaps and bounds, longevity increased in developed nations, and now—now we are actually talking about colonizing Mars! What a species!

It all seemed so wildly successful. It calls to mind a story told by Herodotus about the ancient Greek legislator Solon. Croesus the king showed Solon all his treasures and then asked Solon who he thought must be the happiest of men, presuming that Solon would name Croesus himself. Solon, however, irritated Croesus by naming only dead men. When pressed as to his reasoning, Solon said,

O Croesus, man is completely a thing of chance. To me you appear to be wealthy and king of many men; but I cannot answer the question that you ask me until I know that you have completed the span of your life well. For the one who has great wealth is not at all more fortunate than the one who has only enough for his daily needs, unless fate attend him and, having everything that is fair, he also end his life well.

Croesus did later fall from power and wealth, and he lived out the remainder of his life in thrall to another king.

Modern society, dependent as it is on agriculture, industry, and technology, is like a rich man whose life has not yet ended. Everything seems fine at the moment, but who can be sure it will last? The world and nature move slowly. Our 10,000 years of farming, our 150 years of industry, our few decades of high technology are only brief experiments in the span of geologic time. So far, they seem to have worked. But we have not yet seen, as Solon might put it, whether civilization has completed the span of its life well.

The Life We Live

The signs indicate that we have gone about as far as we can with industry, agriculture, and technology as we know them if we wish to remain healthy. The evidence of our planet's exhaustion is everywhere. We see that exhaustion in the Great Lakes and along our coasts. We see it in the deserts that are creeping across long-overgrazed grasslands. We see it in an atmosphere thickened by the burning of tropical rainforests. We have sought to change natural habitats into products that we can use directly—forests into croplands and lumber, for example. As a result, in large regions of the world we are creating ecological deserts armored with soils baked iron-hard by a tropical sun.

We can look at some hard data, too, and see signs of earthly fatigue. The world houses about 5 billion people, but only about a billion of us live the affluent life-style that one could generically call, regardless of nationality, the American Way of Life. Is it likely that the Earth could survive if all 5 billion of us adopted that life-style? The World-watch Institute, in its *State of the World 1991* report, makes this ominous point:

> Long before all of the world's people could achieve the American dream, however, the planet would be laid waste. The world's 1 billion meat eaters, car drivers, and throwaway consumers are responsible for the lion's share of the damage humans have caused to common global resources. For one thing, supporting the lifestyle of the affluent requires resources from far away. A Dutch person's consumption of food, wood, natural fibers, and other products of the soil involves exploitation of five times as much land outside the country as inside—much of it in the Third World. Industrial nations account for close to two thirds of global use of steel, more than two thirds of aluminum, copper, lead, nickel, tin, and zinc, and three fourths of energy.
>
> Those in the wealthiest fifth of humanity have . . . pumped out two thirds of the greenhouse gases that threaten the earth's climate, and each

◀

Trash floating in the Anacostia River, Washington, D.C., is just another small, obvious sign of an abused environment. Such accumulations of junk indicate a widespread disrespect for an already-stressed planet.

year their energy use releases perhaps three fourths of the sulfur and nitrogen oxides that cause acid rain. Their industries generate most of the world's hazardous chemical wastes, and their air conditioners, aerosol sprays, and factories release almost 90 percent of the chlorofluorocarbons that destroy the earth's protective ozone layer. Clearly, even 1 billion profligate consumers is too much for the earth.

To see how severe are the degradations caused by those 1 billion, we need look at only a single industry—the military—and then imagine the course of environmental history if all nations were affluent enough to bankroll such a beast. Figures compiled by the Worldwatch Institute indicate that the Pentagon is perhaps the single largest oil consumer in the United States, perhaps in the world. Every year it uses enough energy to run all the urban mass transit systems in the United States combined for nearly 14 years. The U.S. military may pump out 150 million tons of carbon dioxide gas yearly. It is probably the nation's largest producer of hazardous wastes, generating in recent years roughly 500,000 tons of toxics annually, more than the top five U.S. chemical companies combined. In 1989, an examination of 1,579 U.S. military bases by the Department of Defense turned up 14,401 cases of toxic contamination. Nearly 100 of them are on the Superfund National Priorities List, and officials at the Environmental Protection Agency believe another 1,000 may be added in the years ahead. The development of nuclear technology has created tons of radioactive plutonium with a half-life of 24,000 years—a legacy for which the twentieth century will still be remembered perhaps half a million years from now, if creatures in that distant time still have memories, and if that distant time still has creatures. The military has done little to clean up its wastes, which have posed threats to human health all over the nation, from Norfolk, Virginia, to Oklahoma City to Denver, Colorado, to towns near the Hanford Reservation in Washington State. Nearly 250,000 people living near Hanford have been exposed to radiation from military nuclear projects, some at a level nearly a million times greater than that considered safe by the federal government.

Clearly, if every nation could afford defense spending the way we can afford defense spending, life on Earth would be precarious indeed. Or think about what would happen if everyone could afford automobiles the way we can. A fifth of U.S. households own three cars, and more than half own at least two. Worldwide, car ownership is considerably less. Only 8 percent of the world's people—about 400 million folks—own cars. But even that small percentage produces an estimated 13 percent of the world's carbon dioxide emissions as well as air pollution and acid rain. Need we even mention the 250,000 people killed in traffic accidents each year? What would happen if every nation adopted the automobile as wholeheartedly as we have in the United States, Canada, and Europe?

It would seem that we in the developed world have adopted a lifestyle that, because of its toxic by-products and its drain on natural resources, simply cannot be long sustained. How can it be sustained when we are poisoning vast food resources, such as the fishery of the Great Lakes? Or when we are draining away the underground water

Overcrowding is evident in all resorts, even Walden Pond, Massachusetts. Overpopulation worldwide and profligate use of world resources by the United States and other first-world countries bode ill for future generations.

reserves that provide the irrigation that makes the growing of wheat and corn possible in the relatively arid Midwest? When the chemical fertilizers we use are poisoning underground drinking supplies in virtually every state? When, according to the Environmental Protection Agency, some 60,000 people die every year from particulate air pollution alone? That's 60,000 lives claimed by an industrial by-product. If an invading army killed six people, we would be up in arms. The president would activate our vast military complex and his voter-approval rating would skyrocket. But when our own industries kill 60,000 of us yearly with one form of air pollution alone, we seem to surrender. Some may call this progress, but how much of this progress can we stand?

Some people say that through technology we can expand without limit. They say that we can meet every environmental challenge with a technological solution. And perhaps we can. Perhaps we can create whatever equipment we need to live through all the environmental problems we cause. We already have the gear needed to survive in a room full of poison gas: filtration masks and bottled oxygen. If we can live in a room full of poison gas, perhaps we can come up with the means to live with water and air pollution. But when we do, will the solution be anything that we want to live with? Is there some point at which the technological solution permits life to go on but to be no longer worth living? If your entire life, for the rest of your years, depended on a gas mask and bottled air, would you be tempted to rip the mask from your face and accept your fate? Is that the tragedy to which technology, untempered by wisdom and scientific knowledge, will bring us?

We are terraforming Earth. We are pumping gases into the atmosphere, just as some scientists hope to do someday on Mars. We are seeding Earth with plants—with a handful of crop species that replace the complex diversity of such natural ecosystems as grasslands and tropical rainforests—just as some of us hope to seed Mars. We are building cities and parking lots and suburbs, all to create a uniquely human habitat. But have we forgotten that the natural Earth, too, is our habitat? That, like the African elephant, we will be unable to survive if we find ourselves suddenly imprisoned in small patches of land—cities and towns—with nothing around us but useless open space, habitats turned into nonhabitats, rainforests into earthenware parking lots and lakes into toxic pools? Can we live without the global habitat that shaped us during millennia of evolution?

Who Gets What's Left?

Many of our political leaders and our industries seem constantly to be scrambling to use up the last of our natural resources. In the Pacific Northwest, loggers want to cut the last of the ancient, virgin forests that survive outside national parks. In Alaska, the oil and gas industry wants to drill on the last pristine stretch of northern coast, which lies in a national wildlife refuge. In the Midwest, farmers want to drain the last of the prairie pothole ponds and suck from the Earth the last of the underground water that they use for irrigation. In Boston Harbor, in the Great Lakes, along our coasts, local and national government agencies want to wait until the last minute before addressing the problem of toxic wastes. In much of the tropics, farmers and loggers want to cut away the last vestiges of the rainforests, those deep dark lands which, in their verdant profusion, seem the very epitome of creation. In Africa, ivory dealers were prevented only by an international trade ban from taking the last of the big tuskers, for which they would undoubtedly have paid a good price, too.

We have used up all but the last morsels of these resources because our forebears—who perhaps can be excused their thoughtlessness because they acted out of scientific ignorance—taught us how to do so. As a result, what we see of the natural world today, however vast it may seem, is only a remnant of the wild places and wild living things that flourished just a few centuries or even a few years ago. In most places we are talking about using up the last 10 percent or the last 5 percent of the land, the grass, the trees, the water. It seems that our economic forces always have some potent reason, some vital need, for laying claim to the last of things, or nearly the last of things, as if just one more bout of exploitation would satiate them. They are like grasping children who cry when told they cannot have all the cookies.

Those who care about wild places and wild things find themselves forced more and more to seek economic reasons for protecting ani-

Canada geese fly to their evening bedding places in Blackwater National Wildlife Refuge, Maryland. Discussions of technical solutions to our environmental woes too often ignore immeasurable aspects of the quality of our lives. Do we wish to merely survive?

mals and wilderness. Ecotourism, we say, may help save the elephant and parts of the forests of Brazil, Belize, Zaire. Constantly, we have to show how nature can be made to satisfy human needs.

But what of the human need for wild places? What of the human need for pleasure and happiness, what of those whose pleasure and happiness derive from the sight of unfettered wild places, of elephants roaming free, of a rainforest dark in shadow, of an arctic plain unsullied by oil wells, of a grassland lush and wild? Somehow the dominant forces in our society, the most powerful economic and political forces, seem to have lost sight of the aesthetic and emotional needs of humankind, perhaps because—and let's hope not—only a few of us have the need for wild places and things.

But what of it? What if only a few of us would feel lifeless without occasional recourse to the natural habitats from which we sprang in some distant past? Relatively few people benefit from the profits of oil and pesticides, but those few have had their share of the natural world and done with it what they would, ignoring our ability to develop alternative materials. The few of us who would like to see some small portions of the globe left untouched—who ask for no profit but pleasure, who derive the greatest benefit from leaving a wild place and its creatures alone—should see our wish granted too. We are not talking about much of a sacrifice on anyone's part. Protected lands in the United States do not comprise even 5 percent of the nation. And besides, the desire to protect the few tracts that are still fairly pristine is, after all, such a harmless wish.

Shall we be angry about the fate of wild places for a paragraph or two? It will not hurt anything or anyone if we are, and it might feel good.

First, we might be angry with our elected officials, those who persistently ignore the need for environmental integrity. We might be angry because they seek to justify environmental destruction in the name of the economy—because it creates jobs and money. We might be angry with their justification because it is ridiculous to say that our economy, perhaps the strongest and most diverse in the world, would collapse if industry were compelled by law to conduct its business without pollution of the air and water. It is *not* too costly to have production without pollution. Again, look at the military. Because it has been allowed to function as if environmental integrity had nothing to do with national defense, we are faced with expensive cleanups of military sites that should not have been polluted in the first place. It will cost us $2 million in nuclear decontamination for every nuclear warhead ever made—roughly $130 billion. It may cost us $40 billion, certainly not less than $20 billion, to take care of the military's toxic wastes. Reducing air and water pollution at our overseas bases will cost us nearly $600 million. We should be angry that such costs have been incurred, along with the health threats they pose, because administrators did not want to pay the initial costs of avoiding these problems. We should resent the need to divert funding toward solving such problems and away from programs such as education and medical care. Every dollar used to correct the ills of profligate industries robs us of something else. Similarly, we should resent every dollar spent on cleaning up the Great Lakes and our coasts, every dollar spent on cleaning up the air, every dollar needed to combat global warming, because each of these problems poses not only an economic threat but also a health threat and because each one of them could have been avoided if business did not wish to remain business-as-usual.

We should be angry for future generations who will be saddled not only with our national debt, but with our environmental debt—the far-flung degradation of the habitat we will bestow upon them. We should be so angry that we say to our political leaders: "Enough stalling. The risks are too great to let you succumb any further to economic forces that focus on profit to the exclusion of everything else—to the exclusion of the quality of life, the beauty of life, and the need for a healthy environment." We should not abide politicians who woo us with the punchlines of fast-food television commercials. We should demand ideas and actions. We should be angry that our political leaders ignore our best scientists, who are telling us we have only 10 years left in which to take action on such problems as global warming. We should be angry that too many of our politicians are playing dice with our future because, they say, it is too costly to engage in a little terraformation here on the home planet.

And we should take that anger and let it cool and use it as a guide, use it to discover the sources and causes of our anger and then use

▶

The McDonald River sweeps away from the Garden Wall mountain face in the background at Glacier National Park, Montana. Nature lovers are being pressed to justify the conservation of "merely" scenic areas by demonstrating their economic value.

that knowledge to determine what must be done to change the conditions that frustrate us. And once we know what must be done, we must do it. Only in taking wise action will we have any hope of finding a way out of the biologically dangerous times in which we live.

What You Can Do

One thing that any citizen can do is simple and easy but, because of our cynicism about elected officials, it seems trite even to suggest it. Nevertheless, you should not underestimate the importance of contacting your elected officials and letting them know how you feel about various legislative measures, such as strong enforcement of the Clean Air Act and the Endangered Species Act. Elected officials often will be swayed if enough people express strong support of a given law. In any event, you can bet that any strong environmental legislation is going to have some heavy hitters batting for the opposition. The automotive industry, for example, has consistently fought, and won, battles against measures designed to save fuel by raising engine-efficiency standards or intended to save human lives by requiring certain safety features in all cars. Similarly, western public-lands ranchers have been strong enough politically to halt any efforts to raise the fees for grazing on public lands. Opponents to environmental reforms will always have their way if you and other staunch supporters of the natural world do not express your opinions to your senators and representatives. Remember that you can effect important changes at the state level as well. Do not forget to drop a line to your state legislators once in a while, too.

Here are a few specific objectives that you can pursue as an environmental activist and a politically and socially concerned citizen:

Support strong Clean Air and Clean Water acts whenever these come up for renewal. These laws can bring manifold benefits. Cleaner air, for example, will save lives, make cities more livable, cut back on greenhouse gases, reduce the level of toxins reaching marine waters and freshwater bodies such as the Great Lakes, and cut back on acid rain. That's a lot of environmental punch for the price of a single letter. For similar reasons, support legislation that seeks strict energy-efficiency standards for motor vehicles and home appliances and that seeks strict controls on industrial effluents going into air or water.

You can help protect rainforests by urging your congressional delegation to support research and protective programs in the tropics. You should also avoid purchasing products made of wood from tropical rainforests. Awareness of the issue is important, so you might want to join a specialty organization such as the Rainforest Alliance in New York City.

Steam escapes from the Trojan Nuclear Power Plant on the Columbia River at Kalama, Washington. Atomic energy is seen as one possible means of improving our environment, but in addition to prohibitive costs the unanswered question remains: What to do with the highly radioactive spent fuel?

An elephant calf raises its trunk to test for danger in Amboseli National Park, Kenya. Protecting pachyderms comes down to nonconsumption of ivory products. If it's made of elephant—tusks, hair, or any other parts—don't buy it.

You can help end the overgrazing of public lands by urging eviction of the livestock that is eating away western grasslands and making our public range one of the regions of the globe most at risk from desertification. Encourage your congressional delegation to seek higher fees for grazing permits or to put a stop to grazing on the public lands altogether. Ask that lands in poor to fair condition be given a chance to recover before grazing continues.

If you are a landlord in the Great Plains, you can take personal action to restore native vegetation by entering some of your land in any of several private, state, and federal land-reserve programs. The federal government has budgeted $1.6 billion for conservation-reserve-program payments, which help landowners to put highly erodible cropland into trees, windbreaks, streamside vegetation, wildlife areas, or grasslands. Another $46 million was approved in permanent easements in the Agricultural Wetlands Reserve. You can find out about these programs by contacting the Soil Conservation Service, the Agricultural Stabilization Conservation Service, the Fish and Wildlife Service, or your state wildlife agency. Those who do not own land can also help by supporting efforts to set aside portions of the Great Plains as parks, national monuments, and other types of protected lands.

The elephant would benefit from increased management funding. Urge your congressional delegation to appropriate the full amount of funding authorized for elephant conservation each year under the African Elephant Conservation Act. Of course, do not buy any products made from elephant materials, such as ivory or skin. Killing for such products threatened to wipe out the African elephant within this decade until trade in ivory and other materials was banned.

Helping out on the ecotourism front is largely a matter of personal commitment. The money you spend in your travels should contribute to wildlife and habitat protection, but be sure that as you travel you take care not to damage local fauna and flora. Otherwise, the funds you leave behind will do little more than pay to make amends. When touring other nations, try to hire local guides and stay at locally run lodges so that the money you spend stays in the host nation.

To help ensure that the management of national parks and other public lands remains in the hands of wildlife and habitat experts, urge your congressional delegation to block any legislation or resolutions that would let political considerations guide such issues as fire suppression on wild lands. Help keep decisions about fire fighting—which is an ecological matter—in the hands of the biological experts.

Above all, do not ever feel defeated. In the conservation field, as in any other, we lose some battles and win others. The important thing to remember is that, regardless of the outcome of any individual battle, the war still goes on. This is especially true on the conservation front. We may have to struggle to save a forest or an arctic plain over and over again, because even if we win the issue one time the developers can always come back another day and try to take over another of the few remaining pieces of wild land. On the other hand, we lose only once, because when a wild place is gone, it is gone for good. Please remember that the only real loss is a loss of heart, a surrender to the forces of destruction. If you can keep your *heart* in the battle, perhaps ultimately you will win your opponent's *mind*.

SUGGESTED READING

General Sources

Adler, Bill Jr. 1991. *The Whole Earth Quiz Book*. William Morrow. New York.

The subtitle asks the question "How well do you know your planet?" Take the quizzes in this book and find out. It'll be fun, too.

Myers, Norman. 1979. *The Sinking Ark: A New Look at the Problem of Disappearing Species*. Pergamon Press. New York.

See next entry, because this book and the next are much alike.

———. 1983. *A Wealth of Wild Species: Storehouse for Human Welfare*. Westview Press. Boulder, Colorado.

This and the previous citation are indispensible because of the tremendous amount of statistics they contain. Numbers numbers numbers. For the same reason, they can be a grind to read. But don't ignore them. They're both getting a bit gray now, but they're still useful.

———. 1991. *The Gaia Atlas of Future Worlds: Challenge and Opportunity in an Age of Change*. Doubleday. New York.

A heavily illustrated, you could even say entertaining, look at the environmental and social problems that plague the world. This short book covers just about everything in a style that readers of *People* magazine will find appealing.

Seager, Joni, ed. 1990. *The State of the Earth Atlas*. Simon & Schuster. New York.

A fascinating collection of maps revealing data on such topics as desertification, global hunger, rainforest destruction, fossil fuel pollution, and automobile use.

Starke, Linda, ed. 1991. *State of the World 1991*. W. W. Norton. New York.

This annual book is one of the finest overviews of the nation's and the world's environmental status. Should be required reading for everyone.

Tropical Rainforests

Denslow, Julie Sloan, and Christine Padoch, eds. 1988. *People of the Tropical Rain Forest*. University of California Press. Berkeley.

Thorough, well-photographed account of an often overlooked rainforest species—humankind.

Hecht, Susanna, and Alexander Cockburn. 1989. *The Fate of the Forest: Developers, Destroyers and Defenders of the Amazon*. Verso. New York.

If you want to know about the Amazon rainforest—anything at all—you'll probably find it here. It's readable, too.

Jacobs, Marius. 1988. *The Tropical Rainforest*. Springer-Verlag. New York.

You will come away from this book with an in-depth knowledge of the biological workings of the rainforest. Probably the best book on the subject, but pretty technical. Probably not for the faint of interest.

Mitchell, Andrew W. 1986. *The Enchanted Canopy: A Journey of Discovery to the Last Unexplored Frontier, the Roof of the World's Rainforests*. Macmillan. New York.

Popular and pleasing account of the subject.

Perry, Donald. 1986. *Life Above the Jungle Floor: A Biologist Explores a Strange and Hidden Treetop World*. Simon & Schuster. New York.

If you start reading this, you will probably find it hard to put down. A very entertaining account of rainforest research by the biologist who developed treetop travel via ropes, pulleys, and guts.

Wildfire

Carrie, Jim. 1989. *Summer of Fire*. Peregrine Smith Books. Salt Lake City, Utah.

Excellent overview of the 1988 fire in Yellowstone National Park. Heavily illustrated with photographs.

Pyne, Stephen J. 1982. *Fire in America: A Cultural History of Wildland and Rural Fire*. Princeton University Press. Princeton, N.J.

Exceedingly thorough coverage of the subject, but deadly dry in the reading. Still, you probably won't find a more complete book on the subject.

———. 1989. *Fire on the Rim: A Firefighter's Season at the Grand Canyon*. Ballantine Books. New York.

More anecdotal than the previous title and a bit less tedious. Great source if you're looking for insights rather than light reading.

Grazing

Dary, David. 1981. *Cowboy Culture: A Saga of Five Centuries*. University Press of Kansas. Lawrence.

Read it just for fun and you'll still learn a lot about a fascinating topic. You may also learn more than you ever wanted to know about the design of the western saddle.

deBuys, William. 1985. *Enchantment and Exploitation: The Life and Hard Times of a New Mexico Mountain Range*. University of New Mexico Press. Albuquerque.

Wonderfully readable account of the history and present environmental problems of a local region.

Sears, Paul B. 1980. *Desert on the March, Fourth Edition*. University of Oklahoma Press. Norman, Oklahoma.

This is an update of a book that first appeared in 1935. It has lasted so long because it's so good. It discusses how human inroads are turning grasslands to deserts.

Great Plains

Anonymous. 1989. *Alternative Agriculture*. National Academy Press. Washington, D.C.

A technical book that provides a great deal of information on agricultural problems and solutions.

Armstrong, Virginia Irving, ed. 1971. *I Have Spoken: American History Through the Voices of the Indians*. Sage Press. Chicago.

A compilation of eloquent speeches, quotes, and observations arranged chronologically from early contact to the present.

Bakeless, John. 1950. *The Eyes of Discovery: The Pageant of North America As Seen by the First Explorers*. J. B. Lippincott Co. Philadelphia.

An excellent book for an examination of pristine America, but probably hard to find.

Black Elk and John G. Neihardt. 1961. *Black Elk Speaks*. University of Nebraska Press. Lincoln.

A deeply moving account of life among the Sioux during the late nineteenth century.

Catlin, George. 1989. *North American Indians*. Penguin Books. London.

Catlin's writing puts the reader back in the West of the early nineteenth century, when buffalo still coursed across the plains, wolves still howled at night under western skies, and Native Americans still followed the ways of their unique cultures.

Ebeling, Walter. 1979. *The Fruited Plain: The Story of American Agriculture*. University of California Press. Berkeley.

A bit dated, but a thorough account of agricultural history, including the development of various crops and machines.

Frazier, Ian. 1989. *Great Plains*. Penguin Books. London.

A best-selling account of the modern plains.

Irving, Washington. 1967. *A Tour of the Prairies*. Pantheon Books. New York.

A skillful writer, Irving unveils the lost West to his readers.

Schlissel, Lillian. 1982. *Women's Diaries of the Westward Movement.* Schocken Books. New York.

A historical perspective on the wagon trains that drove West before the Civil War, including long excerpts from the diaries of women who made the journey.

Oceans

(With special thanks to Carl Safina and Ellen Rulseh of the National Audubon Society)

Anonymous. 1988. *A Citizen's Guide to Plastics in the Ocean: More Than a Litter Problem.* Center for Marine Conservation. Washington, D.C.

Describes the problem and suggests solutions, including methods for personal involvement.

Anonymous. 1990. *Decline of the Sea Turtles: Causes and Prevention.* National Research Council, National Academic Press. Washington, D.C.

Focuses on species in U.S. waters, including their biology, distribution, and mortality. Offers recommendations for improved conservation.

Banister, Keith, and Andrew Campbell. 1985. *The Encyclopedia of Aquatic Life.* Facts on File. New York City.

A general overview and reference work on aquatic life.

Bullock, David K. 1989. *The Wasted Ocean: The Ominous Crisis of Marine Pollution and How to Stop It.* Lyons & Burford, Publishers. New York City.

Discusses pollution problems in different marine habitats and offers solutions.

Carson, Rachel. 1961. *The Sea Around Us.* Signet Science Library Books, The New American Library. New York.

This modern classic discusses the evolution and biology of the ocean, the tides, and the life within.

Edgerton, Lynne T. 1991. *The Rising Tide: Global Warming and World Sea Levels.* Island Press. Washington, D.C.

An overview of how global climate change is likely to affect the ocean.

Hay, John. 1979. *The Run.* W. W. Norton & Co. New York City.

Originally published in 1959, this minor classic chronicles the life of the alewife, a type of herring, off New England.

Holing, Dwight. 1990. *Coastal Alert: Ecosystems, Energy, and Offshore Oil Drilling.* Island Press. Washington, D.C.

Details the meeting of oil development and ocean coasts.

Horton, Tom. 1989. *Bay Country: Reflections on the Chesapeake.* Ticknor & Fields. New York City.

An eloquent and entertaining series of essays about the natural and unnatural history surrounding the bay and its politics.

McGoodwin, James R. 1992. *Crisis in the World's Fisheries: People, Problems, and Politics.* Stanford University Press.

An assessment of worldwide fisheries management problems and processes.

Mowat, Farley. 1984. *Sea of Slaughter.* The Atlantic Monthly Press. Boston.

An overview of the centuries of waste and plunder that European and American fleets brought to the ocean's living resources, including an examination of events leading to the extinction of several species.

Talbot, F. H., and R. E. Stevenson, editors. 1991. *Oceans and Islands.* Smithmark. New York City.

Covers general marine biology, physical oceanography, island biology, and future prospects.

Thorne-Miller, Boyce L., and John G. Catena. 1991. *The Living Ocean: Understanding and Protecting Marine Biodiversity.* Island Press. Washington, D.C.

Focuses on threats while comparing coastal and oceanic marine biodiversity and their conservation and examining international programs, conventions, and laws.

Warner, William W. 1977. *Distant Water: The Fate of the North Atlantic Fisherman.* Little, Brown, and Company. Boston.

A riveting chronicle of the hazardous endeavors undertaken by international commercial fleets in pursuit of marketable fish.

Great Lakes

Ashworth, William. 1986. *The Late, Great Lakes: An Environmental History.* Alfred A. Knopf. New York City.

Everything you could ever want to know about the Great Lakes, from pre-European contact to present-day pollution. As readable as it is thorough.

Colborn, Theodora E., Alex Davidson, Sharon N. Green, R. A. (Tony) Hodge, C. Ian Jackson, and Richard A. Liroff. 1990. *Great Lakes, Great Legacy?* The Conservation Foundation, Washington, D.C., and The Institute for Research on Public Policy, Ottawa, Ontario.

Contains a great deal of up-to-date information and data, but you really have to want to get at it because this book reads like the public-policy document that it is.

Elephant

Bull, Bartle. 1988. *Safari: A Chronicle of Adventure.* Viking. London.

An account that conveys all the adventure and excitement of the original safaris, from their beginnings to now.

Carrington, Richard. 1959. *Elephants: A Short Account of Their Natural History, Evolution, and Influence on Mankind.* Basic Books. New York.

Excellent. The subtitle tells you everything you need to know about this book. A bit dated on the natural history, for which you should see Douglas-Hamilton and Moss, below.

DiSilvestro, Roger L. 1991. *African Elephant: Twilight in Eden.* John Wiley & Sons. New York.

An overview of the evolution, history, and biology of the elephant, with a chapter on the ivory trade. Heavily illustrated with photographs by Page Chichester.

Douglas-Hamilton, Iain and Oria. 1975. *Among the Elephants.* The Viking Press. New York.

The popular, fascinating, and highly readable account of these biologists' first years in the field, told with feeling.

Moss, Cynthia. 1988. *Elephant Memories: Thirteen Years in the Life of an Elephant Family.* Fawcett Columbine. New York.

She keeps a sharp focus on behavior but also discusses the elephant's survival problems.

Scullard, H. H. 1974. *The Elephant in the Greek and Roman World.* Cornell University Press. Ithaca, New York.

Completely fascinating and irreplaceable.

Williams, Heathcote. 1989. *Sacred Elephant.* Jonathan Cape. London.

A lot of photos back up a lot of quotes, which is all this book is. As a compendium of fascinating comments on the elephant, it is quite wonderful to read or just peruse.

Wilson, Derek, and Peter Ayerst. 1976. *White Gold: The Story of African Ivory.* Taplinger Publishing Company. New York.

Quite thorough, quite readable.

Ecotourism

Rinehart, Mary Roberts. 1983. *Through Glacier Park in 1915.* Roberts Rinehart. Boulder, Colorado.

A highly readable and exciting story that takes you back to yesteryear, when the only source of transportation in Glacier National Park was feet—your own or a horse's.

Runte, Alfred. 1990. *Trains of Discovery: Western Railroads and the National Parks.* Roberts Rinehart. Niwot, Colorado

A brief, quick reading account of how turn-of-the-century railroad companies helped shape and support the national park system in the United States.

ILLUSTRATION CREDITS

LOOK HOMEWARD, CONSERVATIONIST

Page 7: © David Julian

CHALLENGE AND CHANGE IN COSTA RICA

Page 16: Kevin Schafer
Page 19: © David Julian/Phototake
Page 21: Kevin Schafer/Martha Hill
Page 23: © David Julian
Pages 26, 28: Gary Braasch © 1991
Page 29: Roger L. DiSilvestro
Page 31: Adapted from F. Gary Stiles and Alexander F. Skutch: *A Guide to the Birds of Costa Rica*.
   Illustrated by Dana Gardner. Copyright © 1989 by Cornell University. Used by permission of the
   publisher, Cornell University Press.
Page 32: Library of Congress
Pages 33, 35: Jack Swenson
Pages 39, 42, 44, 47: Chris Wille/Rainforest Alliance

TRIAL BY FIRE

Page 53: Ted Wood
Page 56: Courtesy Buffalo Bill Historical Center, Cody, Wyoming

WHERE THE DEER AND THE ANTELOPE PLAYED

Pages 76, 105: Page Chichester, with thanks to pilot Jerry Hoogerwerf and Project Lighthawk.
Pages 78, 82: Library of Congress

AS LONG AS THE GRASS SHALL GROW

Pages 106, 121: Ed Pembleton
Page 109: Courtesy Joslyn Art Museum, Omaha, Nebraska
Pages 110, 112, 123, 124: Library of Congress
Pages 119, 126, 130: NEBRASKAland Magazine/Nebraska Game and Parks Commission
Page 127: Ron Klataske
Page 128: Roger DiSilvestro

OUR FATAL SHORES

Page 148: Bradley Clift

SWEETWATER SEAS: AN ILLUSION OF PURITY

Page 164: Robert Michaud/GREMM
Page 169: Courtesy National Archives of Canada
Page 170: Environmental Remote Sensing Center and Sea Grant College Program, University of
   Wisconsin-Madison
Page 173: Buffalo and Erie County Historical Society
Pages 185, 187: Tom Lucas
Page 188: Thomas A. Schneider

HATARI, WATEMBO: AFRICA'S ELEPHANTS AT RISK

Page 201: John Michael Fay
Page 213: Library of Congress
Page 214: Herve Morand
Pages 217, 220: William Thompson

ROADS LESS TRAVELED

Page 235: (bottom) Megan Epler Wood

PLANS FOR PLANETS

Page 262: Reprinted by permission: Tribune Media Services

INDEX